American Shelf and Wall Clocks

A Pictorial History for Collectors

Revised & Expanded 2nd Edition

Robert W.D. Ball

4880 Lower Valley Road, Atglen, PA 19310 USA

DEDICATION

This book is dedicated to all clockmakers, both past and present, who by their knowledge, skills, experimentation and love of the art, have given us these precision instruments of exquisite beauty and superbly graceful line; without their genius, the world would be a lesser place.

Robert W. D. Ball

-Revised price guide 1999
Copyright © 1992 & 1999 by Robert W.D. Ball
Library of Congress Catalog Card Number: 99-63175

Typeset in Swis721 HV BT/Swis721 BT

ISBN: 0-7643-0905-6
Printed in China
1 2 3 4

Published by Schiffer Publishing Ltd.
4880 Lower Valley Road
Atglen, PA 19310
Phone: (610) 593-1777; Fax: (610) 593-2002
E-mail: Schifferbk@aol.com
Please visit our web site catalog at
www.schifferbooks.com

This book may be purchased from the publisher.
Include $3.95 for shipping. Please try your bookstore first.
We are interested in hearing from authors
with book ideas on related subjects.
You may write for a free printed catalog.

In Europe, Schiffer books are distributed by
Bushwood Books
6 Marksbury Avenue
Kew Gardens
Surrey TW9 4JF England
Phone: 44 (0)181 392-8585; Fax: 44 (0)181 392-9876
E-mail: Bushwd@aol.com

ACKNOWLEDGMENTS

I would like to take this opportunity to thank everyone who has made the researching and organization of this book so much easier. Faced with a seemingly overwhelming task, the cooperation received was absolutely fantastic from all concerned. Without this kind of help, my job would have been far more laborious and lengthy, and without the good-humored advice of my dear wife (which I didn't always graciously accept!) format and grammar would have been conspicuously different. She is the peg upon which I hang what abilities I possess.

One of the first persons contacted, *Lindy Larson* of Larson's Clock Shop, Main Street, Westminster, Vermont, 01585, has been a veritable fount of knowledge, experience, and unqualified help. With the aid of his wife Karen, who took the time to transcribe notes, descriptions, and captions, my job was made much simpler than I had any right to expect. Thanks to this book, I now feel that I have two new friends in Vermont.

I also owe a great token of appreciation to all the folks at *Richard A. Bourne Co., Inc.*, Hyannis Port, Massachusetts, 02647. Their help started with Sally Green whos good nature put up with my many calls, and continued to Claudia Bourne who unstintingly gave of her valuable time to make available their vast library of photo files and catalogs going back over the years. Lastly, thanks to Dick Bourne for knowing that knowledge made available is knowledge shared by all.

The professionalism and generosity of all the people who assisted me at Skinner, Inc., 2 Newbury Street, Boston, Massachusetts, 02116 and 357 Main Street, Bolton, Massachusetts, 01740, have been extremely gratifying. Stephen Fletcher has cheerfully answered innumerable questions over the phone, while Alicia Gordon and Sarah Hamill did their very competent best to make my search for material that much easier.

Two former strangers who have now become friends are Joyce Stoffers and Stuart Mitchell of the American Clock and Watch Museum, 100 Maple Street, Bristol, Connecticut, 06010. Without their help, kindliness, consideration, and expertise, this book would have been much more difficult to complete. My thanks, my friends! For those whose interest is piqued by this book, the American Clock and Watch Museum, has over 3,000 clocks and watches dating from 1680 to the present in an historic 1801 house with spacious new wings. Hundreds of clocks are running and striking hourly. Open daily 10-5, March 1 - Nov 30 (closed Thanksgiving). Modest admission. Membership information upon request. Call (203) 583-6070 for more information.

I would also like to thank Linda Stamm and Regina Madigan of Helen Winters Associates, Plainville, Connecticut, for opening their files to me.

Finally, heartfelt thanks go to my valued friend of more years than I can count, Henry Wichmann, of Farmington, Connecticut. Hank is a modest but very knowledgeable collector of clocks; he also has had the long suffering patience to listen to me talk about layouts, captions and descriptions ad infinitum, all the while quietly offering his very valid opinions, which have helped me immeasurably.

If I have left anyone out in my thanks, it has certainly been unintentional...to repeat, everyone with whom I have been in contact has been cooperative in a way I would never have expected before I started this project.

Six variations of shelf and cottage clocks, top to bottom and left to right:
Cottage Clock, Seth Thomas Clock Co., Thomaston, Connecticut, c. 1880. Walnut veneered case with fine original label, 30-hour time-and-strike movement.
Height 9 1/2 in. $300
Shelf Clock, Seth Thomas Clock Co., Thomaston, Connecticut, c. 1880. Rosewood veneered case, 8-day time-and-strike movement.
Height 12 1/2 in. $200
Miniature Cottage Clock, Seth Thomas Clock Co., Thomaston, Connecticut, c. 1880. Mahogany veneered case, 30-hour time-and-strike movement.
Height 9 1/2 in. $175
Cottage Clock, unknown maker, c. 1880. Rosewood veneered case with band inlay, 8-day time-only movement.
Height 11 1/4 in. $100
Cottage Clock, Waterbury Clock Co., Waterbury, Connecticut, c. 1890. Rosewood veneered case, 30-hour time-and-strike movement.
Height 13 1/2 in. $125
Cottage Clock, unknown maker, c. 1875. Solid walnut case, 30-hour time-only movement.
Height 9 3/4 in. $100
Richard A. Bourne Co., Inc.

INTRODUCTION

This book was compiled to give visual pleasure and a historical overview of the various styles of American shelf and wall clocks developed over the centuries. These are the clocks that average citizens were able to afford. From the perspective of today's collector, the majority of the clocks shown in this book can be acquired for relatively fair and modest prices. I have focused on the style of case shown throughout the history of timepieces in America rather than in the particular movement housed within the case, for in choosing a particular clock, the potential owner was more concerned with the way the clock looked than how efficient the movement was in comparison to others.

The flow of emigration to the colonies commencing in the 1600s included clockmakers and timepiece repairers, masters, journeymen and apprentices, who were trained both in the British Isles and on the Continent. At first settling near their points of debarkation in New England and New York City, this group of craftsmen eventually spread to upper New York state, Ohio, Philadelphia and its environs, Richmond, Virginia, Savannah, Georgia and, later, even as far west as St. Louis.

Early clockmakers were frequently men of the general metals trade, since limited demand made other means of supporting themselves and their families necessary. These early colonial artisans were also casters of metal, gunmakers, blacksmiths, jewelers, silversmiths and practitioners of the many arcane arts of metal work. Due to the demand for cases to house the clocks, woodworkers and movement makers soon found themselves in common bond.

By the end of the 1700s, clockmakers were working as millwrights, house builders, joiners and cabinetmakers in conjunction with their work in the metal arts. By this time, the heavier clock cases had become old fashioned, with cases of more graceful proportions — such as the banjo, the pillar and scroll and the column cases — gaining in public appeal. In the rural communities, casemakers attempted to "grain paint" pine cases so as to simulate the more expensive woods. Following the lead of the English, American artisans developed their own small cased, small movement, wall-hung clocks. These improved timepieces led to further experimentation in design, resulting in the lyre, diamond head, and girandole.

As the need for accuracy became a higher priority, the wall chronometers, marine timekeepers, regulators, and precision clocks useful to jewelers and railroad officials appeared. As the drop design was shortened, what we now call school-house clocks became available, and by the middle of the 1800s, the fashionable looking glass clock evolved.

The appeal of shelf or mantel clocks was not only that they were less expensive than clocks of grander proportions, but that they were also able to be displayed more conveniently in the ordinary household. Mass means of production, which came into being around the time of the War of 1812, caused a veritable mushrooming in the sales of all manner of clocks, and it was during this time that the names of Terry, Jerome, and Willard achieved such prominence in the world of timekeeping. The Willards of Massachusetts were responsible for the early development of the shelf clock during the Federal period. Clocks were made to order for the well-to-do, and were understandably expensive and time consuming to complete.

The breakthrough in making clocks affordable to the public at large came when Eli Terry began the mass production of wooden clocks, setting the standards and leading the way for other industrialists to do the same. As the years passed, the melding of man and machine created great strides in the innovative field of clockmaking, combining beauty and grace with mechanical progress. But it was frequently the itinerant peddler who made sure that every household in his territory had a clock to prominently display on their wall or shelf.

Many talented individuals followed in Terry's footsteps, with large companies developing in the Connecticut River region by the middle of the nineteenth century. New mechanical advances, different types of materials, and new ideas all inspired the surge in design, manufacture and marketing of the thousands of clocks that were then being produced.

Men like Olcott Cheney, Seth Thomas, Hiram Camp, Joseph Ives and Silas Hoadley were able to indulge themselves in new areas of design and material, bringing forth the iron front, papier-mâché[1] and porcelain cases in rococo form, as well as the classical Greek, Roman and Egyptian designs so admired during that period.

As might be expected, the more ornate, expensive wall clocks began to be supplanted by the less expensive shelf clock, as borne out by the advertising of the middle 1800s. Even more sadly, though the factories continued to produce, as we entered and progressed through the 20th century, demand steadily fell, and old, proud names disappeared from the scene. Many venerable factories fell silent or were forced to diversify in order to stay in business. Today, replaced in large part by foreign products that are less expensive to produce, the core of our quality timepiece makers has disappeared, probably never to be seen again. It is in tribute to these ingenious, persevering gentlemen of generations past that I have compiled this pictorial history, for they are deserving of our utmost respect and admiration.

Three assorted Cottage clocks, from left to right:
Rare Cottage Clock, T. S. Sperry, New York, c. 1860. Pine case with applied raised decorations, painted transferred tablet, painted zinc dial, maker's label, 30-hour time movement with pendulum suspended from the top of the case.
Height 12 in. $350
Cottage Clock, Henry Sperry & Co., New York, c. 1860. Black painted case with striped decoration, excellent painted tablet with a stenciled eagle, painted zinc dial inscribed by maker, 30-hour time movement with the pendulum suspended from the top board of the clock.
Height 12 in. $250
Scarce Cottage Clock, Chauncey Jerome, Bristol, Connecticut, c. 1850. Mahogany veneered case, painted and transferred tablet, painted wooden dial, 30-hour time movement.
Height 13 in. $250
Richard A. Bourne Co., Inc.

Seven different shelf clocks, from top to bottom and left to right:
Rare Cottage Clock, unsigned, Connecticut, c. 1860. An early cottage clock with a rosewood veneered case with a pattern inlay around the top edge, features a painted zinc dial, chamfered back board, and an unusual time-only movement.
Height 11 in. $275
Scarce Cottage Clock, Terry Clock Co., Waterbury, Connecticut, c. 1860. Mahogany veneered case with painted tablet, painted zinc dial, maker's label and 30-hour "ladder" movement.
Height 11 3/4 in. $150
Scarce Cottage Clock, unsigned, Connecticut, c. 1860. Mahogany veneered case with a very fine and delicate painted tablet, and an unusual Terry-type "ladder" movement with slight variations.
Height 11 1/2 in. $175
Scarce Cottage Clock, Terry Clock Co., Waterbury, Connecticut, c. 1860. Mahogany veneered case with a painted tablet, painted zinc dial with blue corner decorations, maker's label and a 30-hour "ladder" movement.

Height 11 3/4 in. $300
Flat Top Cottage Clock, Seth Thomas Clock Co., Plymouth, Connecticut, c. 1865. Pine Case with walnut veneer, painted zinc dial, 30-hour time-and-strike movement, painted tablet and a maker's label.
Height 14 in. $400
Shelf Clock, One Of The City Series, "St. Louis," Seth Thomas Clock Co., Thomaston, Connecticut, c. 1870. Fine case with intricate moldings veneered with blonde mahogany, transferred floral tablet, painted zinc dial, paper label, 8-day time-and-strike movement inscribed with the maker's name.
Height 15 1/2 in. $400
Cottage Clock, Terry & Andrews, Bristol, Connecticut, c. 1855. Mahogany front case with pine sides and a truly exceptional painted tablet, painted zinc dial, 30-hour time-and-strike movement inscribed by the maker.
Height 12 1/4 in. $300
Richard A. Bourne Co., Inc.

CONTENTS

Clock values vary immensely according to the condition of the clock, the location of the market, and the overall quality of the design and manufacture. All these factors make it impossible to create an absolutely accurate value, but we can offer a guide. The values herein reflect what one could realistically expect to pay at retail or at auction. Neither the author nor the publisher are responsible for any outcomes resulting from consulting the values indicated in this book.

62.4.97

Cottage Clock, Waterbury Clock Co., Waterbury, Connecticut, c. 1875. Cottage case, gilt borders, floral pattern on tablet, green scrollwork on face, 8-day brass movement.
Height 13 1/2 in. $150
Museum of Connecticut History

Small Cottage Clock, Seth Thomas, Plymouth, Connecticut, c. 1837-1844. Whitish grey cottage case, gilt on black geometric pattern on tablet, 30-hour brass "Patent Spring" movement.
Height 14 1/2 in. $200
Museum of Connecticut History

62.4.56

Rosewood Veneer Cottage Clock, Jerome & Co., New Haven, Connecticut, c. 1860. Octagonal top, gilt trim, tablet of flowers and foliage in slight relief, 30-hour brass movement.
Height 13 in. $275
Museum of Connecticut History

Cottage Shelf Clock, Seth Thomas, Thomaston, Connecticut, c. 1868-1880. Gilt cottage clock with small decal on tablet, greenish face, gilt door frame, 8-day wood and brass movement.
Height 15 in. $250
Museum of Connecticut History

Federal Mahogany Pillar and Scroll Clock, Riley Whiting, Winchester, Connecticut, c. 1820. Old finish.
Height 30 in. $900
Robert W. Skinner, Inc.

Chapter One

Shelf Clocks

Pillar and Scroll Shelf Clock, Wadsworth and Turner, Litchfield, Connecticut, c. 1825-1827. Case assembled at Wadsworth and Turner shop, brass finials, original tablet of estate scene with geometric floral border, 30-hour movement by Bishop and Bradley, Watertown, Connecticut. Center finial missing.
Height 29 1/2 in. $1200

Two miniature Pillar and Scroll clocks, from left to right:
Miniature Pillar and Scroll Clock, Chelsea for Black, Star and Gorman, c. 1930. Mint condition example, with reverse painted glass, 8-day time-and-strike balance wheel movement.
Height 24 1/4 in. $700
Miniature Pillar and Scroll Clock, Chelsea for Tiffany & Co., c. 1930. Original and mint, reverse painted glass, 8-day time-and-strike balance wheel movement.
Height 24 1/2 in. $700
Richard A. Bourne Co., Inc.

Pillar and Scroll Shelf Clock, Riley Whiting, Winchester, Connecticut, c. 1830-1835. Pillar and scroll case, brass finials possibly not original, tablet of landscape with floral border, 30-hour wooden movement.
Height 30 3/4 in. $700
Museum of Connecticut History

Rare Upside-Down Pillar and Scroll Clock, Silas Hoadley, Plymouth, Connecticut, c. 1825. A fine mahogany case with scrolled ears, turned columns, cut out feet, superb original painted wood dial, original painted tablet, original upper glass and a rare ivory bushed 30-hour wood "upside-down" movement with original weights. Maker's label has a picture of Benjamin Franklin with his motto, "Time is Money."
Height 28 3/4 in. $3500
Richard A. Bourne Co., Inc.

Rare Pillar and Scroll Clock, probably Pennsylvania, c. 1820. A very unusual large shelf clock with a fine mahogany case, turned feet, turned finials; in the lower section of the door is a hand-painted portrayal of Flowers on board, 8-day time-and-strike movement with a round dial.
Height 48 in. $4800
Richard A. Bourne Co., Inc.

Pillar and Scroll Shelf Clock, Seth Thomas, Plymouth, Connecticut, c. 1822-1827. Terry-type pillar and scroll case, tablet is a copy of an original by Rene Shaples, 30-hour wooden movement. Height 31 1/4 in. $1200
Museum of Connecticut History

Pillar and Scroll Shelf Clock, Eli and Samuel Terry, Plymouth, Connecticut, c. 1824-1827. Brass finials, tablet of large white house near water, 30-hour wooden movement. Height 31 3/4 in. $1000
Museum of Connecticut History

Pillar and Scroll Shelf Clock, Eli Terry, Plymouth, Connecticut, c. 1816. Original painting on glass tablet, 30-hour wooden movement. Height 33 in. $650
Museum of Connecticut History

Pillar and Scroll Shelf clock, Norris North, Torrington, Connecticut, c. 1820. Tablet of house in treed yard, original finials, 30-hour Torrington-type horizontal movement. Height 31 1/2 in. $1000
Museum of Connecticut History

Pillar and Scroll Shelf Clock, Elisha Neal, New Hartford, Connecticut, c. 1830. Tablet of house and seascape, this clock was cased and finished by Neal, incorporating a 30 hour wooden movement made by Samuel terry. Height 32 in. $600
Museum of Connecticut History

Pillar and Scroll Shelf Clock, Mark Leavenworth, Waterbury, Connecticut, c. 1828-1829. Original finials, half columns, flowers in bowl painted on face. Height 31 in. $1250
Museum of Connecticut History

Pillar and Scroll Shelf Clock, Charles Kirke, Bristol, Connecticut, c. 1828-1831. Original tablet of farmhouse, 30-hour wooden movement.
Height 29 1/2 in. $650
Museum of Connecticut History

Pillar and Scroll Shelf Clock, Jerome & Darrow, Bristol, Connecticut, c. 1824-1833. Carved splat, brass finials, 30-hour wooden movement.
Height 31 in. $700
Museum of Connecticut History

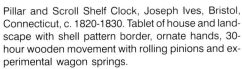

Pillar and Scroll Shelf Clock, Joseph Ives, Bristol, Connecticut, c. 1820-1830. Tablet of house and landscape with shell pattern border, ornate hands, 30-hour wooden movement with rolling pinions and experimental wagon springs.
Height 27 in. $6000
Museum of Connecticut History

Pillar and Scroll Shelf Clock, Joseph Ives, Bristol, Connecticut, c. 1820. Lovely oversize pillar and scroll, brass finials, original top tablet of house and landscape, lower tablet of vase of flowers and green curtains, 8-day brass movement with solid steel plates and wheels, alarm and 1/2 second pendulum.
Height 37 1/2 in. $4200
Museum of Connecticut History

Pillar and Scroll Shelf Clock, Chauncey Ives, Bristol, Connecticut, c. 1835. Small solid brass finials, tablet scene with floral border, 30-hour wooden movement.
Height 35 in. $1300
Museum of Connecticut History

Miniature Pillar and Scroll Shelf Clock, Erastus Hodges, Torrington, Connecticut, c. 1825-1830. Brass finials, original tablet of town square with music academy, 30-hour wooden "Torrington" horizontal movement.
Height 28 3/4 in. $2500
Museum of Connecticut History

Pillar and Scroll Shelf Clock, Silas Hoadley, Plymouth, Connecticut, c. 1825-1840. Pillar and scroll, brass finials, original tablet, 30-hour wooden movement with ivory bushings.
Height 30 3/4 in. $1250
Museum of Connecticut History

Pillar and Scroll Shelf Clock, Ephraim Downes, Bristol, Connecticut, c. 1825-1830. Original finials, lovely original tablet with floral border, 30-hour wooden movement.
Height 31 1/4 in. $1300
Museum of Connecticut History

Pillar and Scroll Shelf Clock, Silas Hoadley, Plymouth, Connecticut, c. 1825-1830. Original brass finials, original tablet, eagle on face, 30-hour upside-down wooden movement.
Height 31 1/4 in. $2000
Museum of Connecticut History

Pillar and Scroll Shelf Clock, Olcott Cheney, Middletown Connecticut, c. 1835. Original brass finials, tablet showing house and landscape scene, 30-hour wooden movement.
Height 31 in. $950
Museum of Connecticut History

Pillar and Scroll Shelf Clock, Lucius B. Bradley, Watertown, Connecticut, c. 1827-1832. Modified Terry design, lower panel fine original reverse painting on glass of Mount Vernon, painted iron dial of Boston manufacture, "Lucius B. Bradley/Watertown/Ct. " on the face, 8-day brass time movement with repeating strike.
Height 31 1/2 in. $850
Museum of Connecticut History

Pillar and Scroll Shelf Clock, Bishop & Bradley, Watertown, Connecticut, c. 1823-1830. Pillar and scroll, reverse painting on glass of windmill and sailboat. 30-hour wooden movement.
Height 28 in. $425
Museum of Connecticut History

Empire Style Shelf Clock, Norris North, Torrington, Connecticut, c. 1825-1831. Small empire stenciled half columns, ship print tablet, 30-hour wooden Torrington-type horizontal movement.
Height 24 in. $850
Museum of Connecticut History

Shelf Clock, Hopkins & Alfred, Harwinton, Connecticut, c. 1840. Stencilled columns and eagle splat, poppy print tablet, 30-hour wooden movement.
Height 33 1/2 in. $900
Museum of Connecticut History

Tall Shelf Clock, Silas Hoadley, Plymouth, Connecticut, c. 1830-1835. Tall stenciled case with carved pineapple finials and paw feet, stenciled columns and eagle splat, original tablets, upside-down 30-hour wooden movement with alarm.
Height 38 1/2 in. $850
Museum of Connecticut History

Shelf Clock, Silas Hoadley, Plymouth, Connecticut, c. 1835. Stenciled quarter columns and splat, carved paw feet and pineapple finials, ship print tablet, 30-hour time-and-alarm movement.
Height 26 1/2 in. $775
Museum of Connecticut History

Double Decker Empire Style Shelf Clock, Eli Terry, Jr., Terryville, Connecticut, c. 1825-1841. Gilt and tortoise gesso half columns, gilt stenciled wooden splat, original tablet, 8-day brass movement with escapement visible in dial cut-out.
Height 31 1/4 in. $900
Museum of Connecticut History

Three half column shelf clocks, from left to right:
Half Column Shelf Clock, Forestville Manufacturing Co., Forestville, Connecticut, c. 1850. Mahogany veneer and stencil decoration, works stamped "W. C. Johnson," excellent dial and glasses, 8-day brass time-and-strike movement.
Height 30 in. $400
Half Column Shelf Clock, Boardman & Wells, Bristol, Connecticut, c. 1830-1840. Excellent glass and dial, stencil decorated half columns and splat, 30-hour brass bushed time-and-strike movement.
Height 32 in. $450
Half Column Shelf Clock, Boardman & Wells, Bristol, Connecticut, c. 1830-1840. Fine label for a Maine dealer in clocks, 30-hour time-and-strike movement.
Height 32 in. $175
Richard A. Bourne Co., Inc.

Three representative shelf clocks, from left to right:
Half Column and Splat Shelf Clock, Samuel Terry, Bristol, Connecticut, c. 1835. 30-hour time-and-strike movement.
Height 34 3/4 in. $200
Half Column Shelf Clock, E. & G. W. Bartholomew, Bristol, Connecticut, c. 1830. Carved columns and eagle splat, 30-hour wooden time-and-strike movement, original label.
Height 34 1/2 in. $300
Half Column and Splat Shelf Clock, Jerome & Darrow, Bristol, Connecticut, c. 1830. Original label, stenciled half columns, 30-hour time-and-strike Groaner movement.
Height 35 in. $325
Richard A. Bourne Co., Inc.

Three representative half column shelf clocks, from left to right:
Half Column Shelf Clock, Orrin Hart, Bristol, Connecticut, c. 1830. Mahogany veneered case, stenciled columns and stenciled splat, 30-hour wood time-and-strike movement.
Height 34 1/2 in. $500
Half Column Shelf Clock, Seymour Hall & Co., Unionville, Connecticut, c. 1830. Eagle splat, marbleized columns, turned feet, 30-hour time-and-strike movement, decal of Alexander Hamilton.
Height 34 in. $450
Half Column Shelf Clock, Riley Whiting, Winsted, Connecticut, c. 1830. Stenciled columns and splat, 30-hour time-and-strike movement.
Height 34 in. $600
Richard A. Bourne Co., Inc.

Two shelf clocks, from left to right:
Shelf Clock, Seth Thomas, Plymouth Hollow, Connecticut, c. 1850. Mahogany veneered case, painted zinc dial, paper label, split gilded columns, extremely choice painted tablet of an eagle on a dark blue background, 30-hour time-and-strike brass weight driven movement retaining iron weights and pendulum.
Height 25 in. $400
Split Column Shelf Clock, Boardman & Wells, Bristol, Connecticut, c. 1835. Mahogany veneered case with stenciled columns and splat, exceptional wood dial, paper label, lithograph in door of "Homeward Bound, the Quay of New York. " 30-hour time-and-strike wood movement with iron weights and pendulum.
Height 34 in. $375
Richard A. Bourne Co., Inc.

Shelf Clock, Marsh, Gilbert & Co., Farmington, Connecticut, c. 1828. Unusually short case for this type of movement, fully carved half columns, fruit basket splat and paw feet, tablet of town square with gilt border, 8-day wooden movement.
Height 31 3/4 in. $850
Museum of Connecticut History

Shelf Clock, Marsh, Gilbert & Co., Farmington, Connecticut, c. 1828. Carved half columns and eagle splat, original tablet of woman, 8-day wooden movement with "improved" ivory bushings.
Height 36 1/2 in. $800
Museum of Connecticut History

Shelf Clock, Norris North, Torrington, Connecticut, c. 1831. Fully carved case with unusually fine quarter columns, fruit basket splat, pineapple finials and paw feet, tablet of classical building, 30-hour Torrington-type horizontal movement.
Height 34 1/2 in. $950
Museum of Connecticut History

Short Case Shelf Clock, Joseph Rogers Bill, Middletown, Connecticut, c. 1840. Carved half columns, splat and paw feet, original tablet of town green with gilt border, 30-hour wooden movement.
Height 27 1/2 in. $875
Museum of Connecticut History

Three desirable shelf clocks, from left to right:
Rare Shelf Clock, Eli Terry Jr. & Co., label, Terry'sville, Ct., 1830. Silas B. Terry, movement attributed. Mahogany case with gold columns, carved eagle crest, exceptional painted tablet, paper label, 8-day time-and-strike "round top" movement with pewter or lead collets and iron weights and pendulum.
Height 36 3/4 in. $950
Cornice and Column Shelf Clock, Birge, Mallory & Co., Bristol, Connecticut, c. 1840. Fine mahogany case with two beveled mirror tablets, enameled wood dial with mirror slide, paper label, 8-day time-and-strike strap brass movement stamped "B. M. & Co.," with iron weights and pendulums.
Height 37 1/4 in. $600
Shelf Clock, John Birge & Co., Bristol, Connecticut, c. 1848. Mahogany with gilded plaster crest, paint decorated columns, painted tablet, exceptional paper label, enameled wood dial, 8-day time-and-strike strap brass movement with iron weights and pendulum. Incorporates the unusual feature of mirror tablets behind the lower columns.
Height 35 in. $750
Richard A. Bourne Co., Inc.

Two choice shelf clocks, from left to right:
Rare Shelf Clock, Birge, Mallory & Co., Bristol, Connecticut, c. 1840. Large mahogany case with choice wood, excellent carved crest, turned columns and ball feet. An original painted tablet in the lower door and two clear glasses in the upper door, original painted wood dial, maker's label and a large strap brass 8-day weight driven time-and-strike movement.
Height 38 1/2 in. $1200
Split Column Shelf Clock, Henry Terry & Co., Plymouth, Connecticut, c. 1830. Fine mahogany case with split columns and carved pineapples, carved crest, black and gold painted wood dial, maker's label and a 30-hour time-and-strike wood movement.
Height 33 3/4 in. $475
Richard A. Bourne Co., Inc.

Cottage Clock, E. N. Welch, Forestville, Connecticut, c. 1860. This case is unusual in that it is grain-painted to resemble exotic woods. Painted zinc dial, reverse painted tablet, 30-hour movement.
Height 12 in. $500
Lindy Larson

Interior view of the Cottage Clock.

Cottage Style Shelf Clock, Brewster & Ingraham, Bristol, Connecticut, c. 1845. Rosewood veneered case with beaded frame, brass spring driven time-and-strike movement.
Height 14 in. $350
American Clock & Watch Museum

Miniature Cottage Clock, S. B. Terry, Bristol, Connecticut, c. 1840. Mahogany case with reverse ogee sides, reverse painted and stenciled upper and lower glasses, painted wood dial, 30-hour time-and-strike movement. Height 14 1/2 in. $1000
Lindy Larson

Interior view of the Miniature Cottage Clock, showing the unusual movement with a center-mounted count wheel; mainsprings are attached behind the movement and anchored to the case by hooks.

Miniature Pillar and Scroll Shelf clock, unsigned, but probably made in Portsmouth, New Hampshire, or the southern Maine area, c. 1820. Mahogany case with turned wood columns, feet and finials, painted iron dial, reverse painted tablet, one-day brass movement with cast lead weight. Height 25 in. $5500
Lindy Larson

Interior view of the Miniature Pillar and Scroll Shelf Clock.
Lindy Larson

Pillar and Scroll Shelf Clock, Norris North, Torrington, Connecticut, c. 1825. Commonly called a "Torrington" style clock because of the unusual features and shape of the movement. Mahogany case with cast brass finials, highly detailed and colorful painted dial and reverse painted tablet with gold leaf border, 30-hour wood movement. Note that the winding arbors are horizontal at "3" and "9. "
Height 31 in. $3000
Lindy Larson

Pillar and Scroll Shelf Clock, Chauncey Ives, Bristol, Connecticut, c. 1825. Mahogany case, painted zinc dial, classically painted tablet, 30-hour wood time-and-strike Terry Style movement.
Height 31 in. $600
American Clock & Watch Museum

Interior view of the Norris North Pillar and Scroll shelf clock. Note the totally different nature of this movement compared to the Terry-style movements.
Lindy Larson

Pillar and Scroll Shelf Clock, Hugh Kearney, Wolcotville, Connecticut, c.
1830. Mahogany veneered case, original painted tablet, painted wood
dial, movement probably purchased from Hopkins & Alfred of Harwinton,
30-hour wood time-and-strike movement.
Height (excluding finials) 28 in. $600
American Clock & Watch Museum

Pillar and Scroll Shelf Clock, Jerome(s') & Darrow, Bristol, Connecticut, c.
1830. Basically a Pillar and Scroll Model, other feet and scrolls were used,
a flat board placed where small pillars would have been so that a half
column would be placed there, reverse crest, mahogany veneered case,
30-hour wood time-and-strike movement.
Height 30 3/4 in. $950
American Clock & Watch Museum

Shelf Clock, C. & N. Jerome, Bristol, Connecticut, c. 1835. Mahogany veneered double decker case, turned half columns, mirror crest, painted center tablet with "Brass Bushed Clock Made By C. & N. Jerome, Bristol, Ct.," lower tablet of ship and eagle, 30-hour brass bushed time-and-strike wood movement.
Height 30 in. $1100
American Clock & Watch Museum

Hollow Column Shelf Clock, E. & G. W. Bartholomew, Bristol, Connecticut, c. 1830. Mahogany case with hollow columns at each side; small diameter cast iron weights travel through the columns to power the 30-hour wood works movement, painted wood dial, mirror in the center sash and reverse painting below.
Height 32 in. $1400
Lindy Larson

Interior view of Hollow Column Shelf Clock.
Lindy Larson

Shelf Clock, Curtiss & Clark, Plymouth, Connecticut, c. 1824. Mahogany veneered case, turned and carved full columns, unusual original painted tablet, 8-day brass time-and-strike movement. One of the first production clocks in America with coiled springs.
Height 22 1/2 in. $850
American Clock & Watch Museum

Hollow Column Shelf Clock, Orin Treadwell, Philadelphia, Pennsylvania, c. 1840. Mahogany case with hollow columns to accommodate elongated round weights, painted zinc dial, cut glass center tablet, replaced lower tablet, 30-hour double weight brass movement is of typical ogee style and is signed "Clarke, Gilbert & Co., Winchester, Ct., U. S. A. "
Height 28 in. $900
Lindy Larson

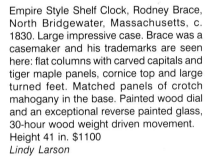

Close-up of the painted tablet in the Rodney Brace Empire Style Shelf Clock.
Lindy Larson

Empire Style Shelf Clock, Rodney Brace, North Bridgewater, Massachusetts, c. 1830. Large impressive case. Brace was a casemaker and his trademarks are seen here: flat columns with carved capitals and tiger maple panels, cornice top and large turned feet. Matched panels of crotch mahogany in the base. Painted wood dial and an exceptional reverse painted glass, 30-hour wood weight driven movement. Height 41 in. $1100
Lindy Larson

Shelf Clock, Boston Clock Co., Boston, Massachusetts, c. 1860. Mahogany veneered case with four turned columns, painted metal dial, etched lyre tablet, brass ornamentation, one year fusee spring driven movement with striking and torsion pendulum. The unique pendulum and striking design enabled clocks to run a full year on one winding.
Height 21 1/4 in. $1200
American Clock & Watch Museum
Page 26

Column and Cornice Shelf Clock, P. Barnes & Co., Bristol, Connecticut, c. 1845. Mahogany veneered two section case, full columns on upper portion with ogee portion below, cornice top, mirror middle tablet, classical building scene in lower tablet, 8-day brass time-and-strike movement like Barnes and Bartholomew & Co., strapped plates, but lantern pinions.
Height 35 3/4 in. $900
American Clock & Watch Museum

Shelf Clock, Rodney Brace, North Bridgewater, Massachusetts, c. 1845. Mirror case, black and gilt gesso half columns, stenciled fruit basket splat, 30-hour wooden time, strike-and-alarm movement.
Height 36 1/2 in. $900
Museum of Connecticut History

Stenciled Column and Splat Clock, c. 1830. Mahogany case with stenciled splat, 3/4 columns, reverse painted tablet, pineapple finials and weight-driven movement. $900

Shelf Clock, M. & E. Blakeslee, Plymouth, Connecticut, c. 1830. Mahogany case with detailed stenciling on columns and crest, carved paw feet, painted wood dial, reverse painted tablet with stenciled border, 30-hour wood time-and-strike movement.
Height 29 in. $1000
Lindy Larson

Interior view of the Blakeslee shelf clock, showing the exceptionally detailed dial.
Lindy Larson

Rare Hollow Column Shelf Clock, Joseph Ives, Bristol, Connecticut, c. 1832. Mahogany case with burl mahogany hollow columns through which the weights drop, carved fruit basket crest and turned feet, round wood dial with mirrored opening, 8-day strap brass movement, unsigned, with pewter winding drums. The label reads "Patent brass eight day clocks manufactured for George Marsh & Co. "
Height 37 in. $6000
Lindy Larson

Shelf Clock, Putnam Bailey, North Goshen, Connecticut, c. 1830. Light mahogany veneer, excluding the sash, carved quarter columns, crest and feet, stenciled and painted tablet, painted metal dial, 30-hour wood time, strike-and-alarm movement, 14 inch pendulum.
Height 24 3/4 in. $1000
American Clock & Watch Museum

Interior view of the Joseph Ives Hollow Column Shelf Clock.
Lindy Larson

Column and Crest Shelf Clock, Chauncey Boardman and Joseph Wells, Bristol, Connecticut, c. 1830. Very unusual design with flat tiger maple columns, carved paw feet and carved eagle crest, three paneled door, painted wood dial with gilded corners, 30-hour time-and-strike weight driven movement with internal alarm.
Height 40 in. $1200
Lindy Larson

Hollow Column Shelf Clock, Chauncey Boardman and Joseph Wells, Bristol, Connecticut, c. 1830. Mahogany case with burl mahogany hollow columns and stenciled crest, painted wood dial, 30-hour wooden weight driven movement.
Height 33 in. $1400
Lindy Larson

Stenciled Column and Splat Shelf Clock, E. Terry & Sons, Bristol, Connecticut, c. 1825-1830. Mahogany case with stenciled splat and quarter columns, reverse painted tablet, 30-hour wood weight driven movement.
Height 28 1/2 in. $1250
Lindy Larson

Carved Column Shelf Clock, Hopkins & Alfred, Harwinton, Connecticut, c. 1825. Very detailed carvings on the columns and crest, elaborate painted wood dial, one-day weight driven wood movement, not of the standard Terry design. Unusual internal alarm.
Height 35 in. $1100
Lindy Larson

Carved Column Shelf Clock, Atkins & Downs, Bristol, Connecticut, c. 1825. Mahogany case with very detailed carving on the columns and crest, painted wood dial, unusual reverse painted tablet with double scenes and stenciled borders, one-day wood movement.
Height 28 1/2 in. $2900
Lindy Larson

Close-up view of the painted tablet in the Atkins & Downs Shelf Clock.
Lindy Larson

Carved Column Shelf Clock, produced in Pennsylvania, c. 1820; labeled by John Conger, Bristol, Connecticut. These cases were produced in Connecticut and sold to the Pennsylvania German craftsmen who installed their own marvelous 8-day brass movements. The painted iron dial and steel hands reflect this Pennsylvania influence, which is also seen in Pillar and Scroll cases. The lower tablet was sometimes painted on tin behind glass. Height 30 in. $4000
Lindy Larson

Interior view of the Carved Column Shelf Clock; note that the cast brass plates are hand filed into elaborate skeletonized designs. 8-day weight driven movement.
Lindy Larson

Close-up view of the Jeromes & Darrow Shelf Clock, showing the most unusual, detailed horse drawn carriage on the splat.
Lindy Larson

Stenciled Column and Splat Shelf Clock, Jeromes & Darrow, Bristol, Connecticut, c. 1825. Full door front with early style bulbous columns, mahogany case, ornate stenciling, painted wood dial with mirror below, 30-hour weight driven wooden "groaner" style movement.
Height 35 in. $600
Lindy Larson

Stenciled Column Clock, Marsh, Gilbert & Co., Farmington, Connecticut, c. 1830. Mahogany case with a very colorful floral dial, reverse painted and stenciled tablet, 30-hour wooden movement.
Height 33 in. $700
Lindy Larson

Column and Splat Clock, Putnam Bailey, Goshen, Connecticut, c. 1830. Mahogany case typical of this maker, with tiger maple quarter columns and scrolled crotch mahogany crest, painted wood dial with mirror below, 30-hour wood weight driven movement.
Height 34 1/2 in. $900
Lindy Larson

Salem Bridge Shelf Clock, R. E. Northrup, New Haven, Connecticut, c. 1834-1836. Mahogany case flanked by flat columns with Ionic capitals, the carved crest extends the full width of the case and is supported from behind by glue blocks, carved paw feet, painted iron dial and reverse painted tablet. 8-day weight driven brass movement.
Height 31 in. $3750
Lindy Larson

Interior view of the Salem Bridge Shelf Clock.
Lindy Larson

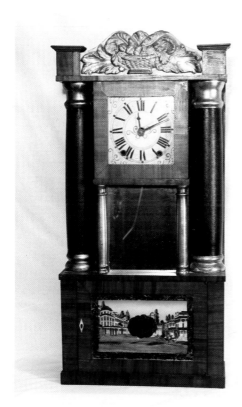

Hollow Column Shelf clock, E. & G. W. Bartholemew, Bristol, Connecticut, c. 1820. Mahogany case with gold leaf columns and crest, painted wood dial, reverse painted tablet.
Height 34 in. $1800
Lindy Larson

Interior view of the Hollow Column Shelf Clock.
Lindy Larson

Detail of the carved crest of the Riley Whiting shelf clock, done in a double cornucopia style.
Lindy Larson

Shelf Clock, Riley Whiting, Winchester, Connecticut, c. 1830. Mahogany case with carved columns and crest, painted wooden dial above and painted tablet below, 30-hour time-and-strike wood works weight driven movement.
Height 35 in. $800
Lindy Larson

Detail of the tablet of the Riley Whiting shelf clock; a combination lithograph and reverse painting. It is a scene of "A View of the North River" and is signed by the firm of D. W. Kellog. This firm also made prints in the manner of Currier and Ives.
Lindy Larson

Carved Column and Crest Shelf Clock, Atkins & Downs for George Mitchell, Bristol, Connecticut, c. 1830. Mahogany veneered case with highly detailed carvings with double set of columns, paw feet and an eagle crest, double doors, wood dial with gilt decorations, 8-day wood time-and-strike weight driven movement.
Height 39 in. $1200
Lindy Larson

Interior view of the carved column and crest shelf clock.
Lindy Larson

Detail of the crest of the clock at left. Note the unusual double eagles in lieu of the more common basket of fruit.

Carved Column and Crest Shelf Clock, Boardman & Wells, Bristol, Connecticut, c. 1830. Made for Hiram Hunt, Bangor, Maine. Beautiful mahogany case, painted and gilded wooden dial with mirror below. 30-hour wood time, strike-and-alarm movement with an alarm wind at the "2;" the alarm bell is hidden behind the carved crest.
Height 34 1/2 in. $750
Lindy Larson

Interior view of the carved column and crest shelf clock.
Lindy Larson

Close-up view of the movement of the carved column and crest shelf clock.
Lindy Larson

Carved Column and Crest Shelf Clock, George Marsh & Co., Farmington, Connecticut, c. 1830. Lovely mahogany case in a double door design, colorful painted dial with mirror below, 8-day wood works movement with ivory bushings; large cast iron weights approximately 9 pounds apiece, with compounded pulleys.
Height 37 in. $1200
Lindy Larson

Interior view of the Marsh, Gilbert & Co. shelf clock.
Lindy Larson

Carved Column and Crest Shelf Clock, Marsh, Gilbert & Co., Farmington, Connecticut, c. 1830. Mahogany case with double doors, carved eagle crest, mirror tablet, 8-day wood works movement with 9 pound cast iron weights and compounded pulleys; weight driven alarm winds at "6" and rings on the same bell as the hour strike.
Height 37 in. $1200
Lindy Larson

Close-up view of the crest and movement of the Marsh, Gilbert & Co. shelf clock.
Lindy Larson

Shelf Clock, Langdon & Jones, Burlington, Connecticut, c. 1834. Mahogany veneered case with carved half columns, crest and feet, painted wood dial, beautiful painted tablet, 30-hour wood time-and-strike weight driven movement.
Height 37 1/2 in. $900
American Clock & Watch Museum

Empire Model Shelf Clock, Marshall & Adams, Seneca Falls, New York, c. 1830. Large mahogany veneered case with cornice top, full columns and hugh paw feet, original painted dial, hands, weights and pendulum, three superb tablets, 8-day brass time-and-strike movement with riveted strap plates.
Height 40 in. $1500
American Clock & Watch Museum

Shelf Clock, unsigned, Connecticut, c. 1840. Mahogany veneered three-tiered case, center mirror tablet with flanking gold leaf full columns, painted tablet in lower sash, carved half columns in upper and lower sashes with carved feet and eagle crest, 30-hour wood time-and-strike movement.
Height 36 in. $850
American Clock & Watch Museum

Empire Style Shelf Clock, Silas B. Terry, Bristol, Connecticut, c. 1840. Mahogany veneered case, 3/4 columns, wood painted dial, original painted tablet, visible escapement, handmade 8-day brass time-and-strike movement, solid backplate and cut-out front plate of 1/16" brass, outside escapement with vertical excapement cock maintaining power with winding arbor of wood with internal rack-and-snail strike. Unusual weight driven rack-and-snail striking movement with pendulum excapement.
Height 28 1/4 in. $700
American Clock & Watch Museum

Close-up of the painted tablet in the Marshall & Adams Empire Style Shelf Clock.
Lindy Larson

Empire Style Shelf Clock, Marshall & Adams, Seneca Falls, New York, c. 1825. Mahogany case; the cornice top is cut to receive a gold leaf and reverse painted tablet of double cornucopia which matches the center panel. The lower tablet is a reverse painting of George Washington, painted wood dial, 8-day double weight brass movement.
Height 40 in. $2000
Lindy Larson

Close-up of the movement in the Marshall & Adams Empire Style Shelf Clock.
Lindy Larson

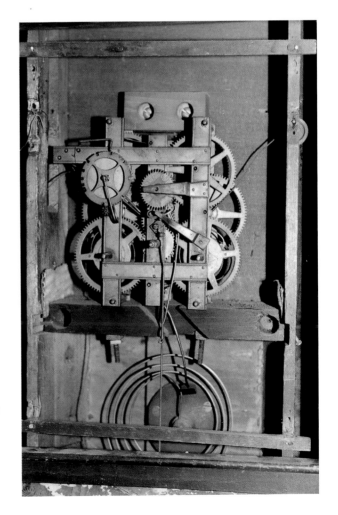

Empire Style Wall Clock, Asa Munger, Auburn, New York, c. 1830. Mahogany case with carved columns and a large overhanging top, the upper glass is backed by a decorated panel of tin, round iron dial is signed, mirrored panel below. This maker's labels are typically small and applied to a wallpapered backboard. Signed 8-day double weight brass movement. Height 39 in. $2500
Lindy Larson

Interior view of the Asa Munger Wall Clock, showing the wallpapered backboard and large weights typical of this maker.
Lindy Larson

"Empire" style Column and Cornice Shelf Clock, Spencer, Wooster & Co., Salem Bridge, Connecticut, c. 1840. Mahogany case with a single door, the bottom partition is removable, fully turned columns with carved capitals, painted wood dial, reverse painted center glass, replacement lower glass, high quality 8-day brass weight driven movement.
Height 33 1/2 in. $1600
Lindy Larson

"Empire" Style Shelf Clock, Seth Thomas, Plymouth Hollow, Connecticut, c. 1860. Rosewood case with octagon half columns, painted zinc dial, two gold leafed reverse painted glasses, 8-day lyre shaped movement is signed and weight driven. This model was also produced as a double dial calendar clock.
Height 31 in. $850
Lindy Larson

"Empire" style Shelf Clock, Philip Smith, Marcellus, New York, c. 1825. Mahogany case with fully turned columns, painted wood dial, replaced lower tablet, 8-day weight driven brass movement.
Height 34 in. $950
Lindy Larson

Interior view of the Philip Smith Empire shelf clock, showing the unusual movement and plate configuration typical of this maker.
Lindy Larson

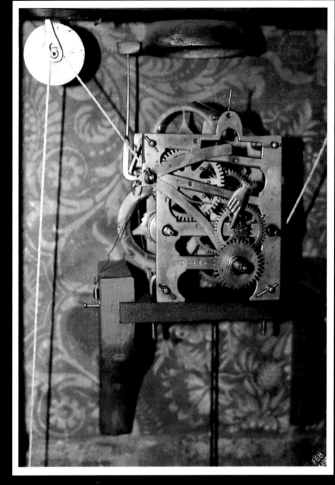

Detail of the movement in the Hotchkiss & Benedict Empire Shelf Clock showing the seatboard arrangement, seconds "hand," interior pulleys and wallpapered backboard which are hallmarks of this maker.
Lindy Larson

Empire Style Shelf Clock, Hotchkiss & Benedict, Auburn, New York, c. 1820. Mahogany case with carved crest, double doors with glass pulls, rounded half columns above and flat columns with carved capitals below, mirrored lower door, very ornate gilded and painted wood dial with double eagle logo. Open seconds bit with brass pointing finger to mark the seconds. Note the upside-down and offset winding holes for the 8-day weight driven movement.
Height 41 in. $2000
Lindy Larson

Double Decker Empire Style Shelf Clock, Hotchkiss & Benedict, Auburn, New York, c. 1820. Figured mahogany case, the lower portion of which has flat applied columns with carved capitals, double doors, very ornately painted wood dial, signed and stamped with a double eagle logo; dial also has seconds bit cut out and the brass hand is actually in the shape of a hand with a pointing finger to mark the seconds. 8-day brass weight driven movement.
Height 37 1/2 in. $1400
Lindy Larson

Interior view of the Hotchkiss & Benedict Empire Shelf Clock. The unusually tall round weights, cast eagle pendulum and small printed label are typical features of this maker.
Lindy Larson

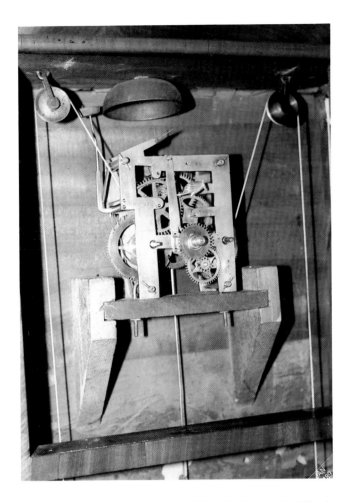

Close-up of the movement in the Hotchkiss & Benedict Empire Shelf Clock.
Lindy Larson

Close view of the Hotchkiss & Benedict label.
Lindy Larson

"Empire Style" Cornice Top Shelf Clock, Birge, Peck & Co., Bristol, Connecticut, c. 1855. Mahogany case, double decker, with upper tablet a depiction of "Barnum's Villa" (Iranistan), and the lower tablet showing J. C. Brown's house, 8-day strap brass weight driven time-and-strike movement. Height 32 1/2 in. $1000
American Clock & Watch Museum

Pillar and Scroll Shelf Clock, Seth Thomas, Plymouth, Connecticut, c. 1822-1827. Terry-type pillar and scroll case, table is a copy of an original by Rene Shaples, 30-hour wooden movement.
Height 31 1/4 in. $1250
Museum of Connecticut History.

Double Decker Shelf clock, C. & N. Jerome, Bristol, Connecticut, c. 1855. Mahogany veneered case, painted metal dial, painted tablet in upper sash, mirror in lower, 8-day time-and-strike brass movement.
Height 37 in. $900
American Clock & Watch Museum

Double Decker Shelf Clock, Richard Ward, Salem Bridge, Connecticut, c. 1841. Mahogany case with cornice top, two glasses in top door, painted metal dial, 8-day rack-and-snail repeating strike, original weights, key, pulleys, pulley covers, hands and bob.
Height 28 in. $950
American Clock & Watch Museum

"Empire Style" Two Tier Shelf clock, M. W. Atkins, Bristol, Connecticut, c. 1850. Mahogany case, 3 inch columns above, divided upper sash, lovely painted and stenciled tablet, mirror in lower sash, painted metal dial, 8-day brass time movement. Height 33 in. $1100
American Clock & Watch Museum

Shelf Clock, Ephraim Downes, Bristol, Connecticut, c. 1835. Mahogany veneered case, paw feet, painted tablet with classical building and foliage, 30-hour wood time-and-strike movement.
Height 31 in. $1000
American Clock & Watch Museum

Shelf Clock, Eli Terry & Sons, Bristol, Connecticut, c. 1830. Mahogany veneered case, unusual flat molded design case without typical columns and splat, painted wood dial, 8-day wood time-and-strike movement.
Height 33 1/2 in. $1100
American Clock & Watch Museum

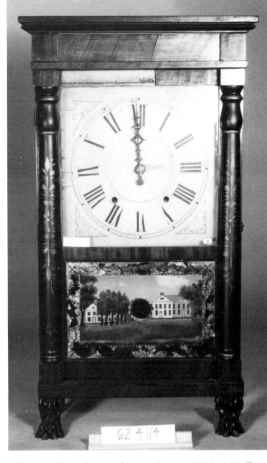

Shelf Clock, Charles Kirke, Bristol, Connecticut, c. 1826-1836. Two panel, square top half column case, columns gilt stenciled, paw feet, original tablet of landscape with classical building.
Height 27 1/2 in. $600
Museum of Connecticut History

Small Empire Style Shelf Clock, Seth Thomas, Thomaston, Connecticut, c. 1868-1880. Full columns, mirror tablet, 8-day brass spring operated movement.
Height 16 in. $200
Museum of Connecticut History

Empire Style Shelf Clock, Eli Terry Jr., Terryville, Connecticut, c. 1835-1841. Large half columns, middle tablet of stylized flowers, lower tablet of landscape, 8-day wooden movement.
Height 36 1/2 in. $800
Museum of Connecticut History

Three Empire shelf clocks, from left to right:
Empire Shelf Clock, Seth Thomas Clock Co., Plymouth, Connecticut, c. 1860. Beautifully veneered rosewood case with two original black and gold glasses, painted zinc dial, maker's label, 8-day time, strike-and-alarm movement.
Height 32 1/2 in. $475
Empire Shelf clock, Seth Thomas Clock Co., Plymouth Hollow, Connecticut, c. 1860. Mahogany veneered case with two painted tablets, painted zinc dial, maker's label and an 8-day time-and-strike brass movement.

Height 32 1/2 in. $425
Empire Shelf Clock, Phillip Smith, Marcellus, New York, c. 1845. Mahogany veneered case with turned columns, painted tablet, mirror, paper label, painted wood dial, and an unusual skeletonized 8-day time-and-strike brass movement.
Height 33 1/2 in. $600
Richard A. Bourne Co., Inc.

Three representative shelf clocks, from left to right:
Empire Double Column Shelf Clock, Terry & Andrews, Bristol, Connecticut, c. 1835-1840. Mahogany case with original label and glasses, plus a good dial; 30-hour brass time-and-strike movement.
Height 26 1/2 in. $250
Empire Shelf Clock, Seth Thomas, Plymouth Hollow, Connecticut, c. 1850. Rosewood case with a good original label, 30-hour brass time-and-strike

movement.
Height 25 in. $175
Empire Double Column Shelf Clock, Sperry & Shaw, New York City, c. 1840. Mahogany case with fine label; 30-hour time-and-strike brass movement. Movement by the Union Clock Co. for Sperry & Shaw.
Height 26 in. $250
Richard A. Bourne Co., Inc.

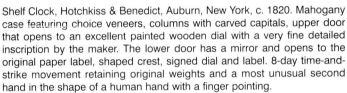

Shelf Clock, Hotchkiss & Benedict, Auburn, New York, c. 1820. Mahogany case featuring choice veneers, columns with carved capitals, upper door that opens to an excellent painted wooden dial with a very fine detailed inscription by the maker. The lower door has a mirror and opens to the original paper label, shaped crest, signed dial and label. 8-day time-and-strike movement retaining original weights and a most unusual second hand in the shape of a human hand with a finger pointing.
Height 38 in. $1450
Richard A. Bourne Co., Inc.

New York Style Shelf Clock, F. C. Andrews, Bristol, Connecticut, c. 1843-1850. Empire style case, black and gilt columns, original tablet, showing American flags and modified Connecticut state seal. Original paper label inside.
Height 26 1/2 in. $475
Museum of Connecticut History

Three different shelf clocks, from left to right:
Miniature Empire Ogee Shelf Clock, Seth Thomas, Thomaston, Connecticut, c. 1870s. Rosewood veneered case, 30-hour brass time-and-strike movement.
Height 15 3/4 in. $200
Empire Shelf Clock, label of Daniel Pratt Jr., Salem, Massachusetts, c. 1830; works by E. C. Brewster, Bristol, Connecticut, c. 1830s. Rosewood veneered case with excellent label, works mounted on a cast iron back

plate, with barrel springs, 30-hour brass spring and works time-and-strike movement.
Height 21 1/2 in. $250
Half-Size Empire Ogee Shelf Clock, Waterbury Clock Company, Waterbury, Connecticut, c. 1870. Mahogany veneered case, 30-hour time, strike-and-alarm movement.
Height 19 in. $250
Richard A. Bourne Co., Inc.

Three examples of Empire style triple decker shelf clocks, from left to right:
Triple Decker Empire Shelf Clock, Forestville Manufacturing Co., Forestville, Connecticut, c. 1840. Landscape and classical building glasses, 8-day time-and-strike weight driven brass movement.
Height 36 in. $400
Triple Decker Empire Shelf clock, Seth Thomas Clock Co., Thomaston, Connecticut, c. 1890. Circular glasses with some flaking, 8-day brass weight driven time-and-strike movement.
Height 32 1/4 in. $250
Triple Decker Empire Shelf Clock, Forestville Manufacturing Co., Forestville, Connecticut, c. 1840. Original glasses with the painting gone unfortunately, 8-day brass time-and-strike movement.
Height 34 3/4 in. $275
Richard A. Bourne Co., Inc.

Extremely Rare Empire Shelf Clock, Joseph Ives, New York, c. 1825. Exceedingly rare clock operated by a lever spring with 8-day time-and-strike movement; the movement itself is very unusual and represents the first in a series of clocks by Ives that utilized this design of patent lever spring. Paper on iron dial inscribed by Joseph Ives, cast pewter bezel, paper label, period pendulum, tablet depicts landscape and classical buildings.
Height 28 1/2 in. $12,500
Richard A. Bourne Co., Inc.

Double Decker Shelf Clock, Henry Terry, Plymouth, Connecticut; probably made at Eli Terry & Sons old factory, c. 1827-1829. Stenciled half columns and splat, carved paw feet, mirror tablet, 8-day wooden movement.
Height 37 in. $550
Museum of Connecticut History

Shelf Clock, Jerome, Thompson & Co., Bristol, Connecticut, c. 1827. Reeded pilaster and scroll, original tablet shows mill with floral border, 30-hour wooden overhead striking "Groaner" movement.
Height 35 in. $800
Museum of Connecticut History

Shelf Clock, Jerome, Darrow & Co., Bristol, Connecticut, c. 1827-1829. Reeded pilaster and scroll, original tablets, the lower one obverse painted on tin, 30-hour wooden "Groaner" movement.
Height 34 in. $250
Museum of Connecticut History

Double Decker Shelf Clock, Lucius B. Bradley, Watertown, Connecticut, c. 1827-1832. Design attributed to Herman Clark, Plymouth, Connecticut, full columns, Boston made painted iron dial, reverse painting of fruit, 8-day brass movement with second hand and repeating strike.
Height 26 in. $1000
Museum of Connecticut History

Double Decker Shelf Clock, Henry C. Smith, Plymouth, Connecticut, c. 1830-1840. Stenciled half columns and splat, center tablet "Henry C. Smith, Bronzer and Painter, Plymouth, Ct.," lower tablet of classical building overlooking pond, 30-hour wooden movement.
Height 32 in. $300
Museum of Connecticut History

Large Double Decker Shelf Clock, Marsh, Gilbert & Co., Farmington, Connecticut, c. 1828. Stenciled half columns and splat, lower tablet of church green, flowers painted on face, 8-day wooden movement by Jerome & Darrow, Bristol, Connecticut.
Height 36 1/2 in. $800
Museum of Connecticut History

Rare Whiting Double Decker Empire Clock, Riley Whiting, Winchester, Connecticut, c. 1830-1835. Half columns, gold scrollwork on dial face, portrait of Jefferson on tablet, 8-day wooden movement with alarm.
Height 36 in. $1000
Museum of Connecticut History

Double Decker Shelf Clock, Atkins & Downs for George Mitchell, Bristol, Connecticut, c. 1831-1832. Carved pineapple half columns, well-carved fruit basket splat, paw feet, original fruit basket stencil and female portrait on tablet, gilt and painted dial, 8-day reverse train movement.
Height 37 1/2 in. $850
Museum of Connecticut History

Double Decker Shelf Clock, C. & N. Jerome, Bristol, Connecticut, c. 1834-1839. Half columns with carved pineapple tops, ornate tablet with ship, 8-day repeating brass movement with lantern pinions.
Height 38 in. $1000
Museum of Connecticut History

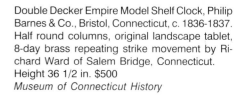

Double Decker Empire Model Shelf Clock, Philip Barnes & Co., Bristol, Connecticut, c. 1836-1837. Half round columns, original landscape tablet, 8-day brass repeating strike movement by Richard Ward of Salem Bridge, Connecticut. Height 36 1/2 in. $500
Museum of Connecticut History

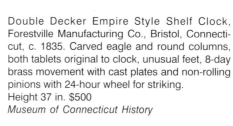

Double Decker Empire Style Shelf Clock, Forestville Manufacturing Co., Bristol, Connecticut, c. 1835. Carved eagle and round columns, both tablets original to clock, unusual feet, 8-day brass movement with cast plates and non-rolling pinions with 24-hour wheel for striking. Height 37 in. $500
Museum of Connecticut History

Double Decker Shelf Clock, Hopkins & Alfred, Harwinton, Connecticut, c. 1831-1842. Extremely rare clock, stenciled half columns and splat, which reads "We Aim To Please," carved paw feet, tablet of classical building, 8-day wooden movement with mahogany plates. Height 36 1/2 in. $1050
Museum of Connecticut History

Double Decker Shelf Clock, Ephraim Downes, Bristol, Connecticut, c. 1845. Carved half columns, splat and paw feet, original tablet of woman, 8-day reverse train wooden movement.
Height 37 1/2 in. $700
Museum of Connecticut History

Double Decker Shelf Clock, Seth Thomas, Plymouth, Connecticut, c. 1837-1844. Carved half columns, eagle splat and paw feet, tablet of village green, floral design on face, 8-day wooden movement with mahogany plates and brass bushings.
Height 37 in. $750
Museum of Connecticut History

Double Decker Shelf Clock, R. and J. B. Terry, Bristol, Connecticut, c. 1836-1856. Gilt and tortoise painted gesso columns, stenciled eagle and military trophy splat, lower tablet "A View of Austria," eagle on center tablet, gilt on white face with black numerals and mirror, 8-day brass, movement with strapped plates and embossed gears.
Height 36 1/2 in. $600
Museum of Connecticut History

Double Decker Shelf Clock, Seth Thomas, Thomaston, Connecticut, c. 1870. Double decker case with gilt half columns, see through works, original tablets, the upper an eagle, flag and shield, the lower, three parrots, 8-day brass movement.
Height 34 in. $650
Museum of Connecticut History

Double Decker Shelf Clock, Forestville Manufacturing Co., Bristol, Connecticut, c. 1839. Gilt gesso fruit basket splat, gilt and painted gesso columns, eagle middle tablet and lower house and landscape tablet original to the clock, improved 8-day brass movement with cast plates and non-rolling pinions with 24-hour count wheel for striking.
Height 37 1/2 in. $700
Museum of Connecticut HistoryST1

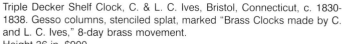

Triple Decker Shelf Clock, C. & L. C. Ives, Bristol, Connecticut, c. 1830-1838. Gesso columns, stenciled splat, marked "Brass Clocks made by C. and L. C. Ives," 8-day brass movement.
Height 36 in. $900
Museum of Connecticut History

Empire Style Triple Decker Shelf Clock, R. Atkins & A. Downs for George Mitchell and R. Atkins, c. 1831-1832. Gesso splat, gilt carved fruit basket, gesso columns, painted tortoise shell, portrait of George Washington in center tablet, gilt ball feet, 8-day wooden reverse train movement.
Height 36 in. $900
Museum of Connecticut History

Unusual Triple tablet Shelf Clock, Jerome & Darrow, Bristol, Connecticut, c. 1827-1833. Fully carved case with fruit basket splat, full length carved half columns, original tablets, unusual short pendulum in tall case makes long weight drop possible, 8-day wooden movement with brass bushings.
Height 41 in. $950
Museum of Connecticut History

Empire Style Triple Decker Shelf Clock, R. Atkins & A. Downs for George Mitchell, Bristol, Connecticut, c. 1831-1832. Carved gilt gesso splat, gilt and painted gesso columns, gilt ball feet, original tablet of Greek revival building, 8-day wooden movement with oak plates by Jerome and Darrow.
Height 37 1/2 in. $875
Museum of Connecticut History

Triple Decker Shelf Clock, Eli Terry, Jr., Terryville, Connecticut, c. 1833-1841. Stenciled half columns and splat, carved paw feet, original tablets, upper portrait of Andrew Jackson, lower, flags and cannons, 8-day wooden movement.
Height 37 in. $800
Museum of Connecticut History

Triple Decker Shelf Clock, Barnes, Bartholomew & Co., Bristol, Connecticut, c. 1834-1835. Gilt and Tortoise gesso columns, carved gilt gesso eagle, gilt paw feet, original tablets, the center tablet of Benjamin Franklin, carved pineapple finials, 8-day riveted strapped brass movement with rolling pinions and unembossed wheels.
Height 37 1/2 in. $950
Museum of Connecticut History

Triple Decker Shelf Clock, Barnes, Bartholomew & Co., Bristol, Connecticut, c. 1833-1836. Fully carved with center mirror, face decorated with gilt and small mirror, lower tablet landscape, 8-day strapped brass movement with lantern pinions and unembossed wheels.
Height 36 1/4 in. $850
Museum of Connecticut History

Triple Decker "Empire" Style Shelf Clock, Chauncey and Noble Jerome, Bristol, Connecticut, c. 1834-1839. Round columns, original tablets, 8-day brass movement with "A" frame and rolling pinions. Height 38 3/4 in. $950
Museum of Connecticut History

Triple Decker Shelf Clock, Forestville Manufacturing Co., Bristol, Connecticut, c. 1835. Gesso columns and eagle splat, center mirror, "Forestville Manufacturing Co. /Bristol. Ct U. S. A. " on face, 8-day brass movement. Height 36 in. $625
Museum of Connecticut History

Triple Decker Shelf Clock, Birge & Peck, Bristol, Connecticut, c. 1830-1856. Gesso columns, eagle splat, ball feet, tablets with reverse painting on glass, with the lower a floral wreath and the middle a geometric design. 8-day brass movement with rolling pinion steel points.
Height 37 1/2 in. $1000
Museum of Connecticut History

Shelf Clock, Jerome & Darrow, Bristol, Connecticut, c. 1827-1833. Large mirror case with stenciled half columns and fruit basket splat, 30-hour wooden movement.
Height 43 in. $625
Museum of Connecticut History

Mirrored Shelf Clock, Marsh, Gilbert & Co., Farmington, Connecticut, c. 1828. Mirror case, stenciled half columns and splat, 30-hour wooden movement.
Height 32 in. $450
Museum of Connecticut History

Double Decker Shelf Clock, E. & G. W. Bartholomew, Bristol, Connecticut, c. 1828-1832. Hollow column case, carved eagle splat, paw feet, acorn finials, center mirror with gilt columns, original glass and tablet, 30-hour wooden movement.
Height 37 1/2 in. $950
Museum of Connecticut History

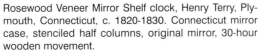

Shelf Clock, Rodney Brace, North Bridgewater, Massachusetts, c. 1845. Mirror case, black and gilt gesso half columns, stenciled fruit basket splat, 30-hour wooden time, strike-and-alarm movement.
Height 36 1/2 in. $450
Museum of Connecticut History

Shelf Clock, T. M. Roverts, Bristol, Connecticut for Elisha Curtis Brewster, c. 1834. Mirror case, stenciled half columns and splat, gilt on face, 30-hour wooden movement.
Height 34 in. $450
Museum of Connecticut History

Rosewood Veneer Mirror Shelf clock, Henry Terry, Plymouth, Connecticut, c. 1820-1830. Connecticut mirror case, stenciled half columns, original mirror, 30-hour wooden movement.
Height 31 1/2 in. $500
Museum of Connecticut History

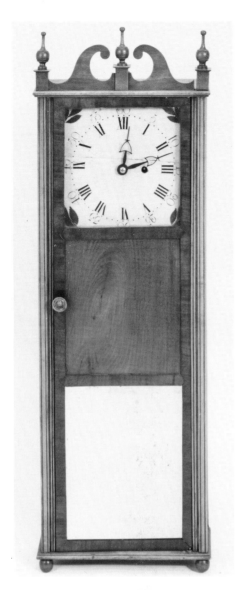

Extremely Rare and Early Shelf Clock, attributed to John Taber, Saco, Maine, c. 1815. Unusual slender design with a birch case and a mahogany veneered door, delicate scrolled crest with turned finials, mirror tablet, painted wood dial and an 8-day time movement with brass bushed iron plates. An interesting feature is that the plates of the movement are held together by long screws that shoulder on the front plate and attach the movement to the case.
Height 37 1/2 in. $2400
Richard A. Bourne Co., Inc.

Three examples of shelf clocks, from left to right:
Half Column and Splat Shelf Clock, Ephraim Downs, Bristol, Connecticut, c. 1830. Columns and splat have fine stencil decoration, excellent label, 30-hour wooden time-and-strike movement.
Height 33 1/2 in. $300
Empire Ogee Shelf Clock, Seth Thomas Clock Co., Thomaston, Connecticut, c. 1870-80. Rosewood veneered case, original label, 8-day time, strike-and-alarm lyre movement.
Height 30 in. $350
Half Column and Splat Shelf Clock, Rollin Atkins, Bristol, Connecticut, c. 1830. Excellent stenciling, label and dial, 30-hour weight driven time-and-strike movement, 42 tooth escapement wheel.
Height 30 1/4 in. $400
Richard A. Bourne Co., Inc.

Triple Decker Shelf Clock, George Marsh & Co., Farmington, Connecticut, c. 1830. Mahogany case flanked by grain-painted columns with gold leaf trim, carved eagle crest, painted wood dial, 8-day weight driven wooden movement with ivory bushings.
Height 39 1/2 in. $1200
Lindy Larson

Miniature Triple Decker Shelf Clock, Birge, Mallory & Co., Bristol, Connecticut, c. 1855. Mahogany veneered triple section case with two painted tablets showing classical scenes, 30-hour brass strap movement with Ives rolling pinions, original bob, weights and dust covers.
Height 25 1/4 in. $1000
American Clock & Watch Museum

Interior view of the George Marsh Triple Decker Shelf Clock.

Triple Tiered Shelf Clock, J. B. Terry, Bristol, Connecticut, c. 1850. Mahogany veneered case, gilt decorated columns, gilt splat, lovely upper painted tablet with mirror lower tablet, 8-day brass time-and-strike movement. A beautiful example in prime condition !
Height 37 in. $1100
American Clock & Watch Museum

Triple Decker Shelf Clock, Birge, Peck & Co., Bristol, Connecticut, c. 1849. Mahogany case with three pairs of columns, the top and bottom pair are half columns, which are grain-painted and gilded, while the center pair are full column and gilded; painted zinc dial, two reverse painted tablets, gilded gesso basket of fruit crest and gilt feet. 8-day weight driven strap brass movement.
Height 36 in. $850
Lindy Larson

View of the movement of the Terry Triple Decker Shelf Clock. Note the unusual rack-and-snail strike peculiar to this Terry product.
Lindy Larson

Triple Decker Shelf Clock, R & J. B. Terry, Bristol, Connecticut, c. 1835. Mahogany case with carved columns above and below, with gilded columns in the center, painted and gilded wood dial, 8-day brass weight driven movement.
Height 37 in. $950
Lindy Larson

Triple Decker Shelf Clock, C. & L. C. Ives, Bristol, Connecticut, c. 1835. Mahogany case with turned mahogany columns and feet, ornately carved crest, painted wood dial has a figure "8" cutout with a small mirror which can be lifted to see or adjust the escapement, 8-day strap brass weight driven movement.
Height 39 in. $750
Lindy Larson

Interior view of the Ives Triple Decker Shelf Clock. Note the "A" frame movement made from brass straps riveted together. Cut-outs in wheels were to conserve brass.
Lindy Larson

Close-up view of the carved crest on the Ives Triple Decker Shelf Clock.
Lindy Larson

Miniature Triple Decker Shelf Clock, Birge, Mallory & Co., Bristol, Connecticut, c. 1840. Mahogany veneered case, gilded and painted crest, columns and feet, brass dial, 30-hour weight driven strap brass movement.
Height 26 1/2 in. $3000
Lindy Larson

Interior view of miniature triple decker shelf clock.
Lindy Larson

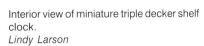

Three Tiered Shelf Clock, C. Jerome & Co., Bristol, Connecticut, c. 1835. Mahogany veneered case, turned half columns above and below, carved caps on upper columns, three painted sashes, the lower with military figure, carved crest, painted wood dial, 8-day square tooth inside rack-and-snail movement, brass solid back plate, cut front plate.
Height 34 in. $1200
American Clock & Watch Museum

"Franklin Alarm" Model Shelf Clock, Silas Hoadley, Plymouth, Connecticut, c. 1835. Mahogany veneered case, half columns with splat, paw feet, lower tablet showing "Time Is Money," unusual inverted, upside-down movement; weight driven alarm is set with the third hand.
Height 37 in. $600
American Clock & Watch Museum

Triple Decker Shelf Clock, Figure "8" Case, C. & L. C. Ives, Bristol, Connecticut, c. 1870. Mahogany veneered case, black painted ornamentation, double scene tablet, 8-day time wagon spring movement.
Height 31 1/2 in. $950
American Clock & Watch Museum

Shelf Clock, Lumas Watson, Cincinnati, Ohio, c. 1830. Imitation paint-grained case, painted wood dial heavily decorated, mirror tablet, 30-hour wood time-and-strike movement.
Height 35 1/2 in. $750
American Clock & Watch Museum

Shelf Clock, Atkins Clock Co., Bristol, Connecticut, c. 1870. Very hand-some case-on-case styling, rosewood with half columns set at a quarter angle, black and gold decorated upper glass, ivory knobs.
Height 18 in. $1450
Lindy Larson

Interior view of the Atkins Clock Co. Shelf Clock, showing the 8-day fusee movement.
Lindy Larson

Pillar and Scroll Shelf Clock, Silas B. Terry, Bristol, Connecticut, c. 1829. Mahogany veneered case, scroll base, stenciled quarter columns, gold gesso crest, eglomise tablet, painted wood dial, 30-hour wooden movement.
Height 28 in. $875
American Clock & Watch Museum

Transition Shelf Clock, Seth Thomas, Plymouth, Connecticut, c. 1828. Mahogany case with carved paw feet, stenciled columns and crest, colorful painted dial, corners decorated with strawberries and blueberries, an unusual design, reverse painted tablet.
Height 29 1/2 in. $1100
Lindy Larson

Transition Model Pillar and Scroll Shelf Clock, Jerome & Darrow, Bristol, Connecticut, c. 1828. This clock is more of a true transition between the pillar and scroll and later stenciled clocks; note the columns are set against a splat and are free standing, giving the sides of the case an indented look. Mahogany case with turned feet, the base having applied molding, gilded dial with interior chapter ring, reverse painted tablet with gold leaf border.
Height 29 1/2 in. $1000
Lindy Larson

Interior view of the Seth Thomas shelf clock.
Lindy Larson

Shelf Clock, Seth Thomas, Plymouth, Connecticut, c. 1828.
Transition-style case in mahogany veneers with carved double cornucopia crest and columns, as well as carved paw feet, colorful painted wood dial with the early style minute markers inside the numerals. Later style dials had the markers outside the numerals. Reverse painting on glass tablet with a stenciled border and pendulum opening, original paper label shows "Invented by Eli Terry, made and sold at Plymouth, Ct. by Seth Thomas. "
Height 29 in. $1200
Lindy Larson

Miniature Ogee Shelf Clock, William S. Johnson, New York, New York, c. 1845. Mahogany case, patriotic reverse painted tablet, painted wood dial, signed 30-hour double fusee movement.
Height 18 in. $700
Lindy Larson

Small Transition Model Shelf Clock, E. & G. W. Bartholemew, Bristol, Connecticut, c. 1830. Mahogany case with carved paw feet, stenciled columns and crest, painted wood dial with gilt decoration, reverse painted tablet, 30-hour wood works weight driven movement.
Height 28 in. $1200
Lindy Larson

Interior view of the Miniature Ogee Shelf Clock, showing the standard movement with the addition of a seat board and feet to hold the double fusees.
Lindy Larson

Very Rare Ogee Clock,
Hills, Goodrich & Co.,
Plainville, Connecticut, c.
1850. Ogee shelf clock
veneered with choice
mahogany, a glazed
door with painted tablet
opening to gilded carved
crest and columns, the
gilded columns with mir-
ror behind. Original dial
and paper label, 8-day
brass time-and-strike
movement with a cast iron
back plate cased in a
brass shell.
Height 31 in. $900
*Richard A. Bourne Co.,
Inc.*

Upside-Down Ogee Style Shelf Clock, Forestville Manufacturing Co., Bristol, Connecticut, c. 1840. Mahogany case with reverse painted tablet, painted zinc dial with unusual center opening. Note the upside-down placement of the winding arbors.
Height 27 in. $500
Lindy Larson

Interior view of the Upside-down Ogee Style shelf Clock, showing the movement. The label notes that the company partners were E. N. Welch, J. C. Brown and N. Pomeroy.
Lindy Larson

Ogee Shelf Clock, Smith & Goodrich, Forestville, Connecticut, c. 1847. Figured mahogany case with painted zinc dial, stenciled geometric glass, rare 30-hour time, strike-and-alarm movement; each arbor is strung to a fusee barrel.
Height 19 in. $750
Lindy Larson

Interior view of the Smith & Goodrich Ogee Shelf Clock, showing the triple fusees mounted to a cast iron bracket. The label notes "Springs with equalized power, warranted not to fail. "
Lindy Larson

Interior view of the E. C. Brewster & Co. Ogee shelf clock.
Lindy Larson

Ogee Shelf Clock, E. C. Brewster & Co., Bristol, Connecticut, c. 1861. Mahogany case, painted and decorated zinc dial with a large opening to view the rather unusual mechanism; 30-hour spring driven movement with a cast iron backplate instead of the usual brass. Solid great wheels with cast iron barrels a part of the backplate.
Height 25 in. $400
Lindy Larson

Ogee Shelf Clock, Birge, Mallory & Co., Bristol, Connecticut, c. 1840. Mahogany case, painted and gilded wood dial with a large center cut so the buyer of the day would know he was getting a quality brass mechanism. 8-day weight driven brass strap movement.
Height 29 1/2 in. $500
Lindy Larson

Interior view of the Birge, Mallory & Co. Ogee shelf clock, showing the movement made by riveting strips of brass together. The label advertises this "Extra Brass" quality.
Lindy Larson

Ogee Shelf Clock, Smith Brothers, New York, c. 1850. Figured mahogany veneer case, unusual reverse painted tablet of state seal of New Jersey. 30-hour weight driven brass movement.
Height 28 in. $500
Lindy Larson

Ogee on Ogee Shelf Clock, Forestville Manufacturing Co., Bristol, Connecticut, c. 1835. Mahogany veneered in an unusual case-on-case style, signed wood dial, stenciled lower glass, 8-day weight driven brass time-and-strike movement.
Height 31 in. $900
Lindy Larson

Interior view of Ogee on Ogee shelf clock.
Lindy Larson

Ogee Shelf clock, C. Jerome, Bristol, Connecticut, c. 1850. Mahogany veneered case, original painted tablet of classical building and foliage scene with spandrels, round brass dial, 30-hour brass movement "#6164" stamped on front plate under cannon arbors.
Height 26 in. $950
American Clock & Watch Museum

Ogee Shelf Clock, Joseph Ives, Bristol, Connecticut, c. 1845. Mahogany veneered case, typical "tin can" mounting with burnished gold columns, painted metal dial, replaced tablet, 8-day time-and-strike roller pinion movement.
Height 30 in. $1100
American Clock & Watch Museum

Rippled Beehive Shelf Clock, J. C. Brown, Forestville, Connecticut, c. 1845. Rosewood case with full-rippled molding on the case and the door, signed zinc dial, cut glass tablet.
Height 19 in. $1800
Lindy Larson

Ogee Double Door Shelf Clock, Forestville Manufacturing Co., Bristol, Connecticut, c. 1846. Mahogany veneered case, etched glass tablet, 8-day time-and-strike weight driven movement.
Height 30 in. $1000
American Clock & Watch Museum

Interior view of the Rippled Beehive Shelf Clock.
Lindy Larson

Interior view of the Full Rippled Beehive Shelf Clock.
Lindy Larson

Full Rippled Beehive Shelf Clock, J. C. Brown, Bristol, Connecticut, c. 1845. Mahogany case with applied full-rippled molding, cut glass tablet, painted zinc dial, 8-day movement with fusees attached to a cast iron bracket. Height 19 in. $1850
Lindy Larson
Rippled Beehive Shelf Clock, J. C. Brown, Bristol, Connecticut, c. 1845. Rosewood case with full-rippled design, cut glass tablet, signed dial. Height 19 in. $1600
Lindy Larson

Interior view of the J. C. Brown Rippled Beehive Shelf Clock, showing the 8-day spring movement; note the pressed brass lyre-shaped ornament which sits above the pendulum.
Lindy Larson

Another view of the J. C. Brown Miniature Rippled Beehive Shelf Clock showing its diminutive proportions beside a standard 19 inch J. C. Brown Beehive.
Lindy Larson

Miniature Rippled Beehive Shelf clock, J. C. Brown, Bristol, Connecticut, c. 1845. Mahogany case with full-rippled frame and door, ivory knob, reverse painted tablet of a wreath of flowers. A very rare model for its height.
Height 15 1/4 in. $2800
Lindy Larson

Interior view of the J. C. Brown Miniature Rippled Beehive Shelf Clock, showing the signed 30-hour movement.
Lindy Larson

Beehive Rippled Front Shelf Clock, J. C. Brown, Bristol, Connecticut, c. 1880. Rosewood veneered case, painted zinc dial, frosted glass tablet, 8-day time-and-strike brass movement.
Height 19 in. $500
American Clock & Watch Museum

Beehive Rippled Front Shelf Clock, Forestville Manufacturing Co., Bristol, Connecticut, c. 1880. Rosewood veneered case, paper label, painted zinc dial, frosted glass tablet, 8-day time-and-strike brass movement.
Height 19 in. $500
American Clock & Watch Museum

Mahogany Veneer Ogee Clock, Atkins & Porter, Bristol, Connecticut, c. 1845. Exceptional glass tablet inscribed with the maker's name, painted zinc dial with colorful corner decorations, paper label, 30-hour time-and-strike movement with cast iron weights.
Height 26 in. $500

On the right is a fine Mahogany and Rosewood Veneered Ogee Clock, J. C. Brown, Bristol, Connecticut, c. 1850. Exceptional tablet, painted wood dial inscribed by the maker, 8-day time-and-strike movement with iron weights and pendulum.
Height 29 in. $400

Richard A. Bourne Co., Inc.

Column Shelf Clock, A. Munger, Auburn, New York, c. 1825-1830. Original eagle pendulum, wallpaper interior, 8-day time-and-strike weight driven brass movement; the second hand is a pointing finger.
Height 38 1/2 in. $1200

Richard A. Bourne Co., Inc.

Three representative Ogee clocks, from the left:
A Mahogany Veneered Ogee Clock, Chauncey Jerome, New Haven, Connecticut, c. 1850 Central painted tablet, exceptional lower tablet, enameled wood dial, exceptional paper label depicting the Jerome Factory, 8-day time-and-strike movement with iron weights and pendulum.
Height 30 in. $600

The middle clock is a scarce Mahogany Veneered Ogee Clock, Eli Terry Jr. & Co., Terryville, Connecticut, c. 1845. Mirror tablet, enameled wood dial, paper label, 8-day time-and-strike "round top" brass attributed to S. B.

Terry with pewter collets, iron weights and pendulum.
Height 33 1/2 in. $350

The right hand clock is a Rosewood Veneered Ogee Clock, Ansonia Brass & Clock Company, Ansonia, Connecticut, c. 1845. Exceptional painted tablet, enameled zinc dial, paper label depicting the Ansonia Clock Factory, 30-hour time-and-strike movement with iron weights and pendulum.
Height 25 3/4 in. $300

Richard A. Bourne Co., Inc.

Three Ogee clocks, from left to right:
Rare Ogee Clock, movement by S. B. Terry, Thomaston, Connecticut, label of Conant & Sperry, c. 1840. In rich mahogany veneers, this clock features a superb original painted tablet with flowers on a blue background in the door; It incorporates an unusual Terry time-and-strike movement with the center arbor mounted below the great wheels, as can be seen by the upside-down winding holes. Typical of S. B. Terry, the escape wheel is unusually large and the great wheels are solid.
Height 27 1/2 in. $850
Scarce Double Door Ogee Clock, Forestville Manufacturing Co., Bristol, Connecticut, c. 1850. A mahogany case with figured veneers, this double door ogee has a fine painted wood dial inscribed by the maker. A painted glass in the lower door, a paper label and an 8-day time-and-strike movement inscribed by the maker that retains the original cast iron weights.
Height 31 in. $850
Scarce Ogee Clock, Birge, Mallory & Co., Bristol, Connecticut, c. 1840. Mahogany veneer case with applied and gilded plaster decorations, a superb etched glass tablet, solid brass dial, and a 30-hour time-and-strike strap movement with roller pinions and original weights. Movement is stamped "B M & Co. "
Height 26 1/4 in. $400
Richard A. Bourne Co., Inc.

Three shelf clocks, from left to right:
Ogee Clock, Elisha Manross, Bristol, Connecticut, c. 1845. Choice mahogany and rosewood veneered case, exceptionally colorful painted tablet, painted zinc dial, paper label and 30-hour time-and-strike movement.
Height 26 in. $300
Ogee Clock, George Marsh, Winchester, Connecticut, c. 1850. Mahogany and rosewood veneered case with an exceptional tablet depicting the "United States Hotel Saratoga," painted wooden dial, paper label, 30-hour time-and-strike brass movement.
Height 25 1/4 in. $375
Ogee Clock, William S. Johnson, New York, c. 1845. Exceptional case with mahogany cross banding and bird's-eye maple veneer, painted tablet, painted wood dial, paper label, 30-hour time-and-strike movement.
Height 27 3/4 in. $350
Richard A. Bourne Co., Inc.

Two Ogee shelf clocks, from left to right:
Rare Ogee Clock, E. C. Brewster, Bristol, Connecticut, c. 1840. Mahogany veneered case, painted tablet, heavy painted zinc dial, and an unusual and rare spring driven movement with a cast iron back plate.
Height 26 in. $450
Rare Double Door Ogee Clock, Jerome & Co., New Haven, Connecticut, c. 1860. Mahogany veneered case, two doors, the lower with an excellent frosted design, painted zinc dial, maker's label and an 8-day time-and-strike weight-driven movement.
Height 30 in. $575
Richard A. Bourne Co., Inc.

Three scarce Ogee clocks, from left to right:
Rare Oversize Ogee Clock, Eli Terry Jr. & Co., Label, Terry's Ville, Ct., 1835; Silas B. Terry, movement. Mahogany veneered case with mirror tablet, exceptional enameled wood dial, paper label, 8-day "round top" movement with heavy plates, pewter or lead collets, iron weights and pendulum.
Height 33 1/2 in. $500
Scarce Oversize Ogee Clock, Philip L. Smith, Marcellus, New York, c. 1830. Mahogany veneered case with a print mounted behind the lower glass, paper label, enameled wood dial, unusual 8-day time-and-strike movement with strap brass plates stamped "216," grid iron pendulum and stamped brass pendulum bob, two iron weights.
Height 36 in. $675
Rare Ogee Shelf Clock, Jacob Jones, Concord, New Hampshire, c. 1820. Unusual version of a New Hampshire Mirror clock with mahogany veneered case, a painted tablet, a mirror tablet, enameled iron dial inscribed "J. Jones, Concord, N. H.," and an 8-day time-and-strike movement with lead weights and pendulum.
Height 31 3/4 in. $600
Richard A. Bourne Co., Inc.

Three different Ogee clocks, from left to right:
Rare Miniature Ogee, The Terry Clock Co., Terryville, Connecticut, c. 1870. This unusual miniature ogee case with rosewood and mahogany veneers was originally constructed to be a weight driven model; this feature was never used, and the company instead used a very unusual movement with round top, strike count wheel on the back of the rear plate, and other minor differences. The movement is inscribed by the maker and the paper label remains on the back of the case. The painted zinc dial and the black and gold tablet are both typical of the Terry Clock Co.
Height 18 in. $600
Miniature Ogee, Seth Thomas Clock Co., Thomaston, Connecticut, c. 1870. Mahogany case with flowing mahogany veneers, painted dial, paper label, fine stenciled glass tablet with the inscription "Seth Thomas Clock Co. " in the background, 8-day time, strike-and-alarm movement.
Height 16 1/2 in. $600
Miniature Ogee, New Haven Clock Co., New Haven, Connecticut, c. 1860. Rosewood veneered case, painted zinc dial, maker's label, fine colorful painted tablet of a beehive, 8-day time-and-strike movement.
Height 18 1/4 in. $350
Richard A. Bourne Co., Inc.

62.4.104

Standard Ogee Shelf Clock, Hiram Welton, Terryville, Connecticut, c. 1841-1845. Standard ogee case, beehive decal with "By Industry We Thrive," original 30-hour brass movement, Eli Terry Patent. Veneer chipped on right front.
Height 22 1/4 in. $450
Museum of Connecticut History

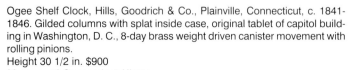

Ogee Shelf Clock, Chauncey and Noble Jerome, Bristol, Connecticut, c. 1834-1839. Half round ogee case, flat top, original tablet with geometric design, reverse painting on glass around face, 30-hour brass movement.
Height 22 in. $325
Museum of Connecticut History

62.4.35

Ogee Shelf Clock, Hills, Goodrich & Co., Plainville, Connecticut, c. 1841-1846. Gilded columns with splat inside case, original tablet of capitol building in Washington, D. C., 8-day brass weight driven canister movement with rolling pinions.
Height 30 1/2 in. $900
Museum of Connecticut History

Ogee Shelf Clock, Hills, Goodrich & Co., Plainville, Connecticut. $925

Ogee Shelf Clock, Silas B. Terry, Terryville, Connecticut, c. 1845-1850. Reverse ogee case with extremely narrow depth, original tablet of capitol at Albany, New York, dial is an engraving printed on paper and applied to wood slating, "Silas B. Terry/Terry'sville Conn" on face, 30-hour brass movement.
Height 24 1/4 in. $575
Museum of Connecticut History

Mahogany and Rosewood Ogee Shelf Clock, Chauncey Boardman, Bristol, Connecticut, c. 1847-1849. Mirror panel, 30-hour brass time, strike-and-alarm movement.
Height 25 1/2 in. $375
Museum of Connecticut History

Standard Ogee Shelf Clock, E. N. Welch, Forestville, Connecticut, c. 1870. Patent design zinc dial, tablet is either transfer or early decal, showing the Merchant Exchange in Philadelphia, Pennsylvania, 8-day brass movement.
Height 28 1/4 in. $300
Museum of Connecticut History

62.4.25

Rosewood Four Column Oriental Style Shelf Clock, Brewster & Ingraham, Bristol, Connecticut, c. 1844-1852. Original acid etched tablet of flower basket, 8-day brass movement with brass springs.
Height 20 in. $900
Museum of Connecticut History

Three rippled shelf clocks, from left to right:
Rippled Beehive Clock, Forestville Manufacturing Co., Bristol, Connecticut, c. 1855. Rosewood veneered case, superb etched glass tablet, exceptional piecrust rippled molding, maker's label, 8-day time-and-strike movement. Height 19 in. $1000
Round Gothic Rippled Front Shelf clock, Forestville Manufacturing Co., Bristol, Connecticut, c. 1850. Fine rosewood veneered case with four turned columns and four finials, a door with a painted glass tablet that opens to the zinc dial inscribed "J. C. Brown, Bristol, Ct. " 8-day time-and-strike movement. Height 19 1/2 in. $1350
Rippled Beehive Clock, Forestville Manufacturing Co., Bristol, Connecticut, c. 1835. Mahogany veneered case with etched glass tablet, painted zinc dial, maker's label, 8-day time-and-strike movement.
Height 19 in. $1000
Richard A. Bourne Co., Inc.

Four variants of shelf clocks, from left to right:
Rare Round Gothic Twin Steeple Shelf Clock, Brewster & Ingraham, Bristol, Connecticut, c. 1845. Rosewood veneered case with a fine painted tablet depicting "The residence of J. C. Brown, Esq., Bristol, Conn. " Painted zinc dial, paper label, 8-day time-and-strike brass movement.
Height 19 3/4 in. $600
Rare Fusee Beehive Clock, Chauncey Boardman, Bristol, Connecticut, c. 1845. Rosewood veneered case with a fine etched tablet, painted zinc dial, 8-day time-and-strike fusee movement inscribed by the maker.
Height 19 in. $500

Beehive Clock, Terry & Andrews, Bristol, Connecticut, c. 1850. Rosewood veneered case with a mirror, fine paper label, painted zinc dial, 8-day time-and-strike lyre movement inscribed by the maker.
Height 19 in. $450
Beehive Clock, Jerome & Co., New Haven, Connecticut, c. 1880. Rosewood veneered case with painted tablet, paper label, painted zinc dial, 8-day time-and-strike movement.
Height 19 1/4 in. $325
Richard A. Bourne Co., Inc.

Two lovely examples of Steeple-on-Steeple Clocks, from the left:
Rare Steeple-on-Steeple Clock with Single Candle, Chauncey Boardman, Bristol, Connecticut, c. 1845. Choice example with two fine original glasses, mahogany veneered case, painted zinc dial, paper label, rare 30-hour time-and-strike movement by Chauncey Boardman with Joseph Ives Patent lever spring, retains period pendulum.
Height 21 3/4 in. $4300
Steeple-on-Steeple Clock, Elisha Manross, Bristol, Connecticut, c. 1840. Fine case veneered vertically with rosewood, two exceptional glasses, paper label, 8-day time-and-strike strap brass fusee movement with pendulum.
Height 24 in. $2100
Richard A. Bourne Co., Inc.

Three examples of steeple clocks are:
On the left, a Sub-Miniature Steeple Clock, Forestville Hardware & Clock Co., Forestville, Connecticut, c. 1853-1855. Mahogany veneered case with frosted glass tablet, delicate turned finials, painted zinc dial, label, 30-hour time movement with pendulum and key.
Height 12 1/4 in. $550
In the middle is a scarce Miniature Steeple Clock, J. C. Brown, Forestville, Connecticut, c. 1850. Rosewood veneered case, painted zinc dial, cut glass tablet, label, 30-hour time, strike-and-alarm movement, with pendulum.
Height 16 in. $475
On the right is a rare Triple Fusee Steeple Clock, Brewster & Ingraham, Bristol, Connecticut, c. 1850-1860. Rosewood veneered case, painted tablet, painted zinc dial, maker's label, 30-hour time, strike-and-alarm fusee movement with pendulum.
Height 20 in. $700
Richard A. Bourne Co., Inc.

Four Steeple Double Decker "Gothic" Shelf Clock, Bristol, Connecticut. Mahogany veneered case, lovely etched and painted glass tablets, with painted zinc dial. $1000

A lovely example of a Mahogany Veneer Double Steeple Shelf Clock, Birge & Fuller, Bristol, Connecticut, c. 1845. Four conical steeples, painted metal dial, two doors with looking glass in middle tablet, frosted lower tablet, four bun feet, Joseph Ives 8-day wagon spring movement.
Height 27 1/4 in. $2100
Robert W. Skinner, Inc.

Rare Double Candlestick Wagon Spring Steeple Clock, Birge & Fuller, Bristol, Connecticut, c. 1845. Fine mahogany veneered case with two exceptional eglomise tablets, painted zinc dial and the rare feature of an 8-day time-and-strike wagon spring movement.
Height 26 in. $2500
Richard A. Bourne Co., Inc.

Three great steeple-on-steeple clocks, from left to right:
Rare Steeple-on-Steeple Clock, Birge & Fuller, Bristol, Connecticut, c. 1845. Mahogany veneered case with four candlesticks, two painted tablets, painted zinc dial, paper label, 8-day time-and-strike fusee movement with pendulum.
Height 26 in. $3000
Steeple-on-Steeple Clock, Terry & Andrews, Bristol, Connecticut, c. 1850. Excellent example with a mahogany and rosewood veneered case, two mint geometric tablets, paper label, painted zinc dial, 8-day time-and-strike lyre movement with original pendulum and pressed brass slide mounted on the pendulum rod above the bob.
Height 26 in. $1900
Fusee Steeple-on-Steeple Clock, Birge & Fuller, Bristol, Connecticut, c. 1850. Mahogany veneered case with two exceptional painted tablets, painted zinc dial, perfect paper label, 8-day time-and-strike fusee movement with pendulum.
Height 26 1/2 in. $2400
Richard A. Bourne Co., Inc.

Three steeple clocks, from left to right:
Scarce Rosewood Veneered Steeple Clock, Terryville Manufacturing Co., Terryville, Connecticut, c. 1860. Nautical tablet with ships, maker's label, 30-hour time-and-strike movement with springs mounted behind back-plate of movement, with pendulum.
Height 20 in. $650
Rare Steeple-On-Steeple Clock, Birge & Fuller Co., Bristol, Connecticut, c. 1845. A very fine example with exceptional tablets, mahogany veneered case, painted zinc dial, over-pasted paper label of J. J. Beals, 8-day time-and-strike fusee movement with pendulum.
Height 26 in. $500
Mahogany Veneered Fusee Steeple Clock, Chauncey Boardman, Bristol, Connecticut, c. 1845. Frosted glass tablet, painted zinc dial, paper label, 30-hour time, strike-and-alarm triple fusee movement with pendulum.
Height 20 in. $500
Richard A. Bourne Co., Inc.

Three fine steeple clocks, from left to right:
Rare Fusee Steeple Clock, Chauncey Jerome, New Haven, Connecticut, c. 1860. Mahogany veneered case, painted zinc dial inscribed "C. Jerome, New Haven, Ct.," and a yellow maker's label. It features a rare 30-hour time-and-strike fusee movement with the pendulum suspended from the top of the case.
Height 19 3/4 in. $450
Steeple Clock, Terryville Manufacturing Co., Terryville, Connecticut, c. 1850. Rosewood veneered case, excellent glass tablet of a Naval engagement, maker's label and a 30-hour time-and-alarm movement with the unusual feature of springs that are mounted behind the back plate of the movement.
Height 19 1/2 in. $1500
Rare Steeple Clock, Brewster & Ingraham, Bristol, Connecticut, c. 1850. Rosewood veneered case, excellent painted tablet, painted zinc dial, choice maker's label, 30-hour time-and-strike movement retaining the original brass springs.
Height 19 3/4 in. $600
Richard A. Bourne Co., Inc.

Three representative steeple clocks, from left to right:
Steeple Clock, Brewster Manufacturing Co., Bristol, Connecticut, c. 1860. Mahogany veneered case with an exceptional painted glass of "Merchant's Exchange, Philadelphia," painted zinc dial inscribed by the maker, paper label, 30-hour time-and-strike movement inscribed "Brewster & Ingraham. "
Height 19 3/4 in. $375
Steeple Clock, Chauncey Jerome, New Haven, Connecticut, c. 1860. Rosewood veneered case, beautifully painted glass tablet, painted zinc dial inscribed by the maker, paper maker's label, 30-hour time, strike-and-alarm, an unusual feature of this clock is that the movement rests on a seat board.
Height 20 in. $350
Steeple Clock, New Haven Clock Co., New haven, Connecticut, c. 1860. Rosewood veneered case with a painted tablet, paper label and an 8-day time-and-strike movement.
Height 20 in. $300
Richard A. Bourne Co., Inc.

Three examples of Steeple clocks, from left to right:
Rare Fusee Steeple Clock with a Rippled Door, Forestville Manufacturing Co., Bristol, Connecticut, c. 1850. In nearly untouched condition, this clock has a mahogany veneered case, exceptional painted glass, painted zinc dial with the maker's name in the arch, and an 8-day fusee time-and-strike movement marked by the maker.
Height 19 3/4 in. $1350
Double Steeple Clock, maker unknown, c. 1850-1860. The mahogany veneered case retains two excellent tablets, one painted and one etched, a made-up 30 day time movement powered by a lever spring intended to be a copy of an Ives Patent. The lever spring was probably added to this case, but is period.
Height 26 in. $1000
Rare Miniature Steeple Clock, Crosby & Vosburgh, New York, c. 1860. A beautiful little steeple clock with a superb painted tablet of birds at a fountain, painted zinc dial, maker's label, 30-hour time movement. Crosby & Vosburgh were the successors to the Jerome Manufacturing Co.
Height 13 1/2 in. $500
Richard A. Bourne Co., Inc.

Rare Double Steeple Clock, Birge & Fuller, with Joseph Ives patent lever, Bristol, Connecticut, c. 1846. Veneered with choice mahogany, this double steeple retains two very fine original cut glass tablets and dial, resting on button feet. The lower door opens to the patent lever and perfect paper label, while the upper door opens to a fine original dial; very rare 8-day time-and-strike spring lever movement.
Height 27 1/4 in. $5000
Richard A. Bourne Co., Inc.

From left to right:
Rare Fusee Round Gothic Shelf Clock, Chauncey Boardman, Bristol, Connecticut, c. 1874. Mahogany veneered case with turned columns and finials, painted zinc dial, etched glass tablet and 8-day time, strike-and-alarm fusee movement inscribed by the maker and "Patented Jan. 1847. "
Height 20 in. $1300
Steeple-On-Steeple Clock, John Birge, Bristol, Connecticut, c. 1850. Mahogany veneered case with zinc dial, paper label, two painted tablets with geometric designs, 8-day time-and-strike fusee movement with pendulum.
Height 26 1/2 in. $1900
Round Gothic Shelf Clock, Brewster & Ingraham, Bristol, Connecticut, c. 1850. Unique case design with flat veneered fronts of the case sides, rosewood veneered case, exceptional etched glass tablet, painted zinc dial, paper label, 8-day time-and-strike movement with rack-and-snail strike and period pendulum.
Height 20 1/2 in. $1300
Richard A. Bourne Co., Inc.

Six fine examples of various styles of clocks, from left to right:
Scarce Fusee Steeple Clock, Boardman & Wells, Bristol, Connecticut, c. 1825. Mahogany veneered case, exceptional painted tablet depicting "Windsor Castle," paper label, painted zinc dial, 30-hour time-and-strike fusee movement.
Height 20 in. $400
Rare Rippled Beehive Clock, J. C. Brown/Forestville Manufacturing Co., Forestville, Connecticut, c. 1860. Fully rippled mahogany case with frosted glass tablet, painted zinc dial inscribed by J. C. Brown, paper label, 8-day time-and-strike movement with pendulum.
Height 19 in. $900
Fusee Steeple Clock, Chauncey Jerome, New haven, Connecticut, c. 1850. Mahogany veneered case, frosted glass tablet, painted zinc dial, paper label, 8-day time, strike-and-alarm triple fusee movement with pendulum.
Height 20 in. $550
Rare Rippled Steeple Clock, J. C. Brown/Forestville Manufacturing Co., Forestville, Connecticut, c. 1860. Mahogany veneered case with applied

rippled moldings, painted glass tablet depicting Mount Vernon, paper label, 8-day time-and-strike movement with pendulum.
Height 20 in. $1600
Steeple Clock, Terry & Andrews, Bristol, Connecticut, c. 1850. Mahogany veneered case with excellent frosted glass tablet, painted zinc dial, paper label, 8-day time-and-strike lyre movement stamped by the maker, brass springs and period pendulum.
Height 20 in. $750
Scarce Fusee Steeple Clock, J. C. Brown/Forestville Manufacturing Co., Forestville, Connecticut, c. 1860. Mahogany and rosewood veneered case with rippled door, exceptional glass tablet, painted zinc dial inscribed by J. C. Brown, paper label, 8-day time-and-strike fusee movement with pendulum.
Height 20 in. $850
Richard A. Bourne Co., Inc.

Three choice shelf clocks, from left to right:
Rare Steeple Clock, Brewster & Ingraham (with Kirk's Patent Iron back plate), Bristol, Connecticut, c. 1844. Fine mahogany veneered case with original etched glass tablet, featuring an 8-day time-and-strike repeating movement with cast iron back plate, with the gong mounted on a base in the shape of a Harp. Movement inscribed by the maker. An interesting feature is the stamp of E. C. Brewster & Co., and the date 1844, on the back of the dial.
Height 20 in. $650
Scarce Four Column Gothic Shelf Clock, Brewster & Ingraham, Bristol, Connecticut, c. 1850. Fine rosewood veneered case with a painted dial, paper label, and an 8-day time-and-strike repeating movement.
Height 17 3/4 in. $950
Rare Steeple clock, Chauncey Boardman (labeled by William S. Johnson), Bristol, Connecticut, c. 1847. Beautiful mahogany veneered case, with the original eglomise tablet, painted zinc dial, paper label and a 30-hour time-and-strike fusee movement inscribed by Chauncey Boardman and dated 1847.
Height 19 1/2 in. $650
Richard A. Bourne Co., Inc.

Double Steeple Shelf Clock, Birge & Fuller, Bristol, Connecticut, c. 1844-1847. J. Ives Patent wooden case, original tablet of flowers and leaves, 30-hour brass wagon spring movement.
Height 24 in. $2500
Museum of Connecticut History

Four Column Gothic Steeple Clock, Brewster & Ingraham, Bristol, Connecticut, c. 1844-1852. Case designed by Elias Ingraham, original acid etched tablet of garland and flower pattern, "Brewster & Ingraham/original/Bristol, Ct. U. S. " on the face, 8-day brass movement with brass springs.
Height 19 in. $1250
Museum of Connecticut History

Double Steeple Shelf Clock, Elisha Manross, Bristol, Connecticut, c. 1845-1854. Original tablets, upper, side wheel boat and dock, lower, sailing ships in harbor, 8-day spring driven fusee movement.
Height 24 in. $1050
Museum of Connecticut History

Rosewood Veneer Steeple Shelf Clock, Chauncey Boardman & J. A. Wells, Bristol, Connecticut, c. 1845. Original tablet decal of steamship on painted background, 30-hour brass movement.
Height 19 1/2 in. $750
Museum of Connecticut History

Steeple Shelf Clock, Chauncey Boardman, Bristol, Connecticut, c. 1847-1849. Steeple clock with original tablet of bluebirds and fountain, 30-hour fusee movement.
Height 20 in. $600
Museum of Connecticut History

Steeple Shelf Clock, Forestville Manufacturing Co., Bristol, Connecticut, c. 1850-1855. Rippled molding, original etched glass tablet, "J. C. Brown/ Bristol, Ct. " on face, 8-day brass movement with fusee.
Height 19 1/4 in. $800
Museum of Connecticut History

Rosewood Veneer Steeple Shelf Clock, Ansonia Clock Company, Ansonia, Connecticut, c. 1851-57. Replacement tablet with geometric design, original paper label.
Height 20 in. $400
Museum of Connecticut History

Miniature Steeple Clock, E. N. Welch Manufacturing Company, Forestville, Connecticut, c. 1875. Rosewood veneer case, painted floral decal tablet, 30-hour brass time-and-strike movement.
Height 14 1/2 in. $375
Museum of Connecticut History

Opposite page Bottom:
Three different shelf clocks, from left to right:
Cast Iron Shelf Clock, Terry & Andrews, Bristol, Connecticut, c. 1851. Mother-of-pearl inlay and painted decoration, brass 8-day time-and-strike movement marked by the maker, and an embossed metal dial marked "Ansonia CLock Co. " Ansonia continued using Terry & Andrews movements after absorbing their parts, and put their name on the dial.
Height 14 in. $200
Cast Iron Shelf Clock, American Clock Co., New York, c. 1860. Iron case with a painted scene, 30-hour balance wheel movement.
Height 14 in. $200
Iron Clock, movement by Noah Pomeroy, case labeled by Bristol Brass & Copper Co., New York, c. 1860. Iron case with painted scene and 30-hour balance wheel movement.
Height 14 in. $150
Richard A. Bourne Co., Inc.

Very Rare Double Steeple Clock, Birge & Fuller, Bristol, Connecticut, c. 1845. Mahogany veneered case with four steeples and two doors, both retaining original glass tablets; the lower door opens to a fine paper label and fusee cones, and the upper door opens to a fine original dial. Rare 8-day brass time-and-strike fusee driven movement. Early repair label, dated 1879, of Leander Freeman, Willimantic, Connecticut, on the back of the case.
Height 26 3/4 in. $2100
Richard A. Bourne Co., Inc.

Standard Acorn Style Shelf Clock, J. C. Brown, Forestville, Connecticut, c. 1840-1850. Door and side arms of laminated wood, tablet shows the signing of the Declaration of Independence, 8-day brass fusee movement.
Height 24 in. $4000
Museum of Connecticut History

Two shelf clocks, from left to right:
Iron Front Shelf Clock, American Clock Co., New York, c. 1860. Unusually large model with painted scenes and floral decorations, maker's label, paper on zinc dial, 8-day time-and-strike movement.
Height 19 3/4 in. $200
Large Iron Front Shelf Clock, unsigned, Connecticut, c. 1865. Unique for its size, this Gothic iron clock has grained and floral painted decorations and mother-of-pearl inlay, painted zinc dial, 8-day time-and-strike movement.
Height 27 1/2 in. $200
Richard A. Bourne Co., Inc.

Below:
Four ornate shelf clocks, from left to right:
Rare Iron Front Mantel Clock, Terryville Manufacturing Co., Terryville, Connecticut, c. 1850. Small iron case with mother-of-pearl inlay and painted gold decoration, painted zinc dial, paper label, unusual 30-hour time-and-strike movement featuring S. B. Terry's patented Torsion balance, and marked "Oct 5th 1852. "
Height 13 in. $400
Iron Front Mantel Clock, unsigned, Connecticut, c. 1860. Ebonized Gothic case with mother-of-pearl decoration and gold designs, painted zinc dial, 8-day time-and-strike movement.
Height 15 1/2 in. $250

Iron Front Shelf Clock, Waterbury Clock Co., Waterbury, Connecticut, c. 1860. Mother-of-pearl inlay and painted decoration, maker's label, paper on zinc dial, 30-hour time, strike-and-alarm movement.
Height 15 1/4 in. $175
Shelf Clock, Daniel Pratt & Sons, Reading, Massachusetts, c. 1860. Iron front case with painted floral and scrolled designs, paper on zinc dial, maker's label, 30-hour time-and-strike movement.
Height 16 3/4 in. $200
Richard A. Bourne Co., Inc.

Three different shelf clocks, from left to right:
Small Iron Front Shelf Clock, unsigned, Connecticut, c. 1860. Mother-of-pearl inlay, painted decorations, painted zinc dial, 30-hour time movement. Height 12 in. $195
Papier-m>chJ Shelf Clock, unsigned, c. 1855. Exceptional case with very fine gold decoration, flowers and mother-of-pearl inlay, painted zinc dial, 30-hour balance wheel time-and-strike movement. Height 13 1/4 in. $225
Miniature Iron Shelf Clock, unsigned, c. 1860. Painted decorations, mother-of-pearl inlay, paper on zinc dial, 30-hour balance movement. Height 10 in. $150
Richard A. Bourne Co., Inc.

Three iron shelf clocks, from left to right:
Gothic Iron Front Shelf Clock, unsigned, Connecticut or New york, c. 1860. Very fine painted case including a scene in the center section with mother-of-pearl inlay, painted zinc dial, 8-day time-and-strike movement.
Height 20 1/2 in. $300
Very fancy Cast Iron Shelf Clock, unsigned, c. 1870. Exceptional painted case with mother-of-pearl inlay, paper on zinc dial, 8-day time-and-strike movement.
Height 18 1/4 in. $300
Iron Front Shelf clock, American Clock Co., New York, c. 1855. Painted floral decorations, mother-of-pearl inlay, paper on zinc dial, 30-hour time, strike-and-alarm movement signed by the maker.
Height 16 1/4 in. $250
Richard A. Bourne Co., Inc.

Three shelf clocks, from left to right:
Cast Iron Shelf Clock, G. Goodrich, Forestville, Connecticut, c. 1860. Mother-of-pearl inlay, painted decoration, painted zinc dial inscribed by the maker, 8-day time-and-strike movement.
Height 16 1/4 in. $325
Papier-m>chJ Shelf Clock, unsigned, c. 1855. Wooden case with papier-m>chJ front painted black with elaborate mother-of-pearl inlay and gold decoration, 8-day time-and-strike movement.
Height 16 1/2 in. $350
Iron Shelf Clock, unsigned, c. 1850. Black case with gold decorations and inlaid mother-of-pearl, painted zinc dial, 30-hour balance wheel movement.
Height 12 in. $300
Richard A. Bourne Co., Inc.

Four Steepled "Oriental Gothic" Rippled Front Shelf Clock, Brewster & Ingraham, Bristol, Connecticut, c. 1850. Rosewood veneered case, fine frosted glass tablet, paper on zinc dial, 8-day brass time-and-strike movement.
Height 20 in. $600
American Clock & Watch Museum

Extremely Rare Sharp Gothic Shelf Clock, J. A. Wells, Bristol, Connecticut, c. 1845. Very unusual case with an ogee molded edge veneered with rosewood; door with a fine original etched eagle tablet opening to an original dial, with Joseph Ives lever spring in the bottom and a fine paper label of J. A. Wells. Standard 30-hour Jerome-type brass time-and-strike movement.
Height 26 1/4 in. $4000
Richard A. Bourne Co., Inc.

Very rare Double Steeple Clock, Birge & Fuller, Bristol, Connecticut, with Joseph Ives patent spring lever, c. 1850. Rosewood veneered case on four button feet, with four steeples and two doors, the upper one with a painted tablet opening to a fine original dial and the lower door with a perfect original tablet opening to the patent spring lever of Joseph Ives. 30-hour brass time-and-strike movement.
Height 24 1/4 in. $3800
Richard A. Bourne Co., Inc.

Steeple Shelf Clock, H. Welton, Terryville, Connecticut, 1840-1845. Mahogany case, frosted glass, very rare weight driven timepiece, runs 27 hours on one winding with a 23/4 pound weight. Compounded pulley. Height 193/4 in. $3200
Lindy Larson

Interior view of the Welton Steeple Clock.
Lindy Larson

Very Rare 3/4 size Steeple Clock, Brewster & Ingraham, Bristol, Connecticut, c. 1845. Mahogany case, gilded feet and finials, painted zinc dial, replaced tablet, unusual 30-hour brass movement with brass main springs. Height 19 3/4 in. $800
Lindy Larson

Interior view of the 3/4 Steeple Clock, showing the movement by Charles Kirk. It has the same front plate as was used in the cast iron backplate models.
Lindy Larson

Steeple Clock, E. C. Brewster & Son, Bristol, Connecticut, c. 1855. Mahogany case with turned finials, signed zinc dial, cut glass tablet, 8-day spring movement.
Height 19 in. $550
Lindy Larson

Steeple-on-Steeple Shelf Clock, E. N. Welch, Forestville, Connecticut, c. 1847. Mahogany case-on-case design with four turned finials, painted zinc dial, two reverse painted tablets, 8-day spring movement. This case style was also made in fusee and wagon-spring models by various companies.
Height 23 in. $1500
Lindy Larson

Interior view of the E. C. Brewster & Son Steeple Clock, showing the signed ribbed movement and pressed brass ornament which sits above the pendulum.
Lindy Larson

Interior view of the Welton Steeple Clock.
Lindy Larson

Rippled Steeple Shelf Clock, J. C. Brown, Bristol, Connecticut, c. 1845. Rippled rosewood case with beveled door, ivory knob, geometric reverse painted glass, 8-day spring movement with separate alarm movement. Height 19 in. $1850
Lindy Larson

Interior view of the J. C. Brown Ripple Steeple Shelf clock.
Lindy Larson

Steeple-on-Steeple Shelf Clock, Birge & Fuller, Bristol, Connecticut, c. 1845. Mahogany case with turned bun feet, two cut glass tablets, over-pasted label of Daniel Pratt, Jr., Boston, Massachusetts. Height 27 1/2 in. $4600
Lindy Larson

Interior view of the Birge & Fuller Steeple-on-Steeple Shelf Clock, showing the 8-day movement powered by wagon-spring mechanism.

Close-up of the movement in the Birge & Fuller Steeple-on-Steeple Shelf Clock, showing the separate brass feet which are riveted to the bottom of the movement and attached to the seatboard. The movement is signed just under the hour post.

Rippled Steeple Shelf Clock, J. C. Brown, Bristol, Connecticut, c. 1845. Rosewood case with beveled door and turned finials, ivory knob, one of the many variations of rippled molding.
Height 19 in. $1900
Lindy Larson

Steeple Clock, New Haven Clock Co., New Haven, Connecticut, c. 1870. Rosewood case with turned finials, painted zinc dial and reverse painted tablet, 8-day spring driven time-and-strike movement.
Height 19 in. $500
Lindy Larson

Interior view of the J. C. Brown Rippled Steeple Clock, showing the typical 8-day time-and-strike movement with separate alarm movement.
Lindy Larson

Interior view of the New Haven Clock Co. Steeple CLock, showing a typical 8-day movement of the period.
Linday Larson

Experimental Steeple Clock, Joseph Ives, Bristol, Connecticut, c. 1831. Mahogany veneered case, painted brass dial, 8-day marine movement; a very unique type of experimental case and movement.
Height 20 in. $1100
American Clock & Watch Museum

Interior view of experimental steeple clock.
American Clock & Watch Museum

Steeple Shelf Clock, Seth Thomas Clock Co., Thomaston, Connecticut. c. 1885. Exquisite walnut veneered case with elaborate fretwork and scrolled features, four steeples, painted metal dial, 8-day brass time-and-strike movement. The attached tag says that case plans are available at $. 25.
Height 24 in. $600
American Clock & Watch Museum

Steeple Clock, Silas B. Terry, Terryville, Connecticut, c. 1847. Mahogany veneered case, painted metal dial, painted tablet, 30-hour brass time-and-strike movement. Fusee spring driven movement has Eli Terry's patent balance wheel escapement.
Height 22 in. $900
American Clock & Watch Museum

Steeple Shelf Clock, Seth Thomas, Plymouth Hollow, Connecticut, c. 1850. Mahogany case of unusual design; the peaked top frames a basket of fruit which is done in gold leaf as are the finials, painted zinc dial, reverse painted geometric tablet, 30-hour weight driven brass movement.
Height 29 in. $900
Lindy Larson

Double Steeple Shelf Clock with Candle Spires, Birge & Fuller, Bristol, Connecticut, c. 1845. Mahogany veneered case, upper and lower sash contain beautifully painted tablets, painted metal dial, 8-day brass roller pinion with wooden fusee.
Height 26 in. $1000
American Clock & Watch Museum

Double Steeple Double Decker Shelf Clock, Birge & Fuller, Bristol, Connecticut, c. 1845. Mahogany veneered case, painted upper and lower tablets, paper on zinc dial, 30-hour wagon spring chain and wheel drive.
Height 24 1/2 in. $1100
American Clock & Watch Museum

Four Steeple Double Decker "Gothic" Shelf clock, Birge & Fuller, Bristol, Connecticut, c. 1845. Mahogany veneered case, attractively etched glass tablets, painted zinc dial, wagon spring movement.
Height 27 1/2 in. $1200
American Clock & Watch Museum

Interior shot of the Birge & Fuller Steeple Clock, clearly showing the wagon spring movement.
American Clock & Watch Museum

Sharp Gothic Steeple Clock, Brewster & Ingraham, Bristol, Connecticut, c. 1840. Mahogany veneered case with 3/4 inch stiles, gold leaf ball feet and finials, metal dial, nice painted tablet, 30-hour brass time-and-strike movement.
Height 20 in. $500
American Clock & Watch Museum

"Patti" Shelf Clock, Welch, Spring & Co., Forestville, Connecticut, c. 1890. There is no model name for this clock because they are very uncommon, not labeled and so far have not been found listed in any of the company catalogs. Walnut case with gold leaf decorated front glass and glass sides, black flocked backboard, fancy brass and sandwich glass pendulum, painted zinc dial with typical "Patti" style hands.
Height 18 in. $1800
Lindy Larson

"Khedive" Model Shelf Clock, Welch, Spring & Co., Forestville, Connecticut, c. 1890. One of the "Patti" series; this model has some variations: the solid walnut case has full columns and a cut glass panel to view the brass pendulum, the double sunk porcelain dial is signed and has a jeweled open escapement, cast brass bezel with beveled glass and an 8-day "Patti" style movement with cathedral gong.
Height 17 1/2 in. $1850
Lindy Larson

Interior view of the Welch, Spring & Co. "Patti" shelf clock, showing the distinctive 8-day "Patti" movement.
Lindy Larson

"Patti" V. P. 68 Model Shelf Clock, Welch, Spring & Co., Bristol, Connecticut, c. 1879. Rosewood veneered case, ornate turned finials and side columns, painted tablet with aperture for viewing the pendulum, 8-day brass alarm spring movement.
Height 19 in. $600
American Clock & Watch Museum

"Cary" Shelf Clock, Welch, Spring & Co., Forestville, Connecticut, c. 1879. One of several models of the "Patti" series. High quality rosewood case of fine workmanship, applied turnings and finials, gold leaf decorated front glass, glass sides, black flocked backboard, beautiful brass pendulum with sandwich glass center, 8-day time-and-strike "Patti" movement.
Height 20 in. $1550
Lindy Larson

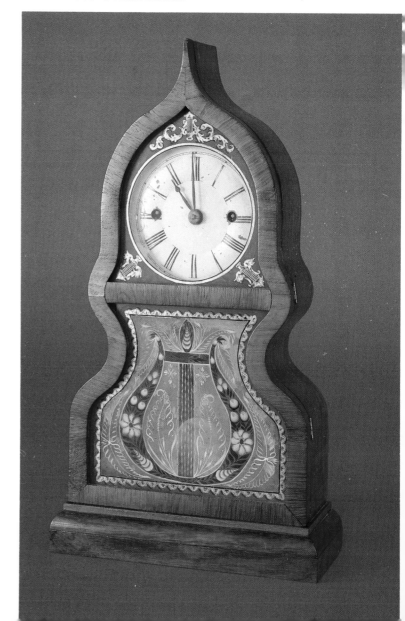

Very Rare Acorn Clock, Forestville Manufacturing Co., Bristol, Connecticut, c. 1849. Laminated case with rosewood veneer, outstanding tablet and decorations, signed on the dial, fine original paper label, 8-day brass time-and-strike movement with fusees.
Height 19 1/4 in. $6500
Richard A. Bourne Co., Inc.

"Acorn" Model Shelf Clock, J. C. Brown, Bristol, Connecticut, c. 1850. Mahogany veneered case, painted zinc dial, black, gold, green and red tablet around dial, lower tablet shows Hartford State House; time-and-strike wood barrel fusee spring movement.
Height 25 in. $1600
American Clock & Watch Museum

Iron Front Mother-of-Pearl Shelf Clock, E. Ingraham & Co., Bristol, Connecticut, c. 1853. Beautifully decorated with gold trim and mother-of-pearl inlay, painted zinc dial, exceptional tablet, 8-day brass time-and-strike spring driven movement.
Height 23 1/2 in. $475
American Clock & Watch Museum

Painted Iron Front Shelf Clock, Ingraham & Stedman, Bristol, Connecticut, c. 1850. Lovely painted case, with painted dial, 8-day time-and-strike movement.
Height 18 in. $500
American Clock & Watch Museum

Interior view of the "Register" double dial calendar clock, showing the calendar mechanism.
Lindy Larson

"Register" Double Dial Calendar Clock, Jerome & Co., New Haven, Connecticut, c. 1860. Solid walnut case with carved crest and interior, signed paper dials, 8-day movement. This clock was also made as a wall model.
Height 27 in. $1450
Lindy Larson

"3-1/2 Parlour" Model Double Dial Calendar Clock, Ithaca Calendar Clock Co., Ithaca, New York, c. 1875. Walnut case with carved and ebonized wood trim, black paper upper dial and rollers, calendar dial and pendulum both made of glass and signed with the Ithaca logo, silvered hands, 8-day movement with bell strike. There were three models made with slight variations.
Height 20 in. $4800
Lindy Larson

Rear view of the "3-1/2 Parlour" Double Dial Calendar Clock.
Lindy Larson

Double Dial Calendar Shelf Clock, National Calendar Clock Co., New Haven, Connecticut, c. 1860. Case is grain-painted to simulate rosewood, painted columns with gilded capitals, signed black dials with brass hands, 8-day movement.
Height 27 in. $1300
Lindy Larson

Calendar Clock, Seth Thomas Clock Co., Thomaston, Connecticut, c. 1880. Mahogany veneered case, painted metal dials, 8-day time-and-strike movement.
Height 43 in. $750
American Clock & Watch Museum

Universal Time Clock, New Haven Clock co., New Haven, Connecticut, c. 1887. Black walnut case, carved, full sash, 8-day brass time-and-strike as the prime mover in the lower dial, contrate gears and extension shaft connect universal time dial showing time in global locations. Experimental model, believed to be a prototype, but never put into production.
Height 36 in. $1250
American Clock & Watch Museum

"Empire" Style Shelf Clock, William Tuller, Hartford, Connecticut, c. 1835. Mahogany veneered case (case by Samuel Camp), basically a three wheel clock, two-part spring barrel and two main springs, with a 4-day brass time-and-strike movement.
Height 18 3/4 in. $995
American Clock & Watch Museum

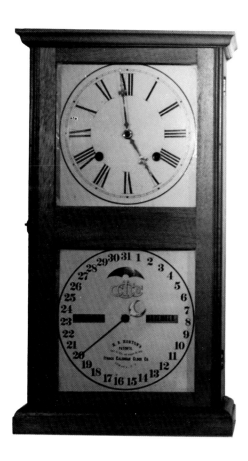

Two attractive calendar shelf clocks, from left to right:
Calendar Clock, Ithaca Calendar Clock Co., Ithaca, New York, c. 1870. Simple rosewood veneered case, painted metal dial, 8-day brass time-and-strike movement, H. B. Horton's patent.
Height 21 in. $500
Calendar Clock, Ithaca Calendar Clock Co., Ithaca, New York, c. 1875. Ornate rosewood veneered case, carved splat and front ebonized, miniature columns, with medallions on both sides, black faced silver dials, crystal pendulum, 8-day time-and-strike movement.
Height 20 1/2 in. $600
American Clock & Watch Museum

Two calendar shelf clocks, from left to right:
Calendar Clock, Jerome & Co., Bristol, Connecticut, c. 1880. Mahogany veneered case, painted metal dial, carved splat and frame, 8-day time-and-strike movement.
Height 33 in. $500
Calendar Clock, Waterbury Clock Co., Waterbury, Connecticut, c. 1880. Pressed oak case, full columns, painted metal dial, 8-day time-and-strike movement. Excellent condition.
Height 29 in. $575
American Clock & Watch Museum

Two good examples of calendar shelf clocks, from left to right:
Calendar Clock, unsigned, Connecticut, c. 1890. Rosewood veneered case, painted metal dial, 8-day time-and-strike movement.
Height 27 in. $500
Calendar Clock, Southern Calendar Clock Co., St. Louis, Missouri, c. 1885. Mahogany veneered case, painted metal dial, "Fashion" tablet, turned finials, full columns, 8-day time-and-strike movement. Excellent condition.
Height 32 in. $600
American Clock & Watch Museum

Three fine calendar clocks, from left to right:
Calendar Clock, unsigned, Connecticut, c. 1885. Arched rosewood case, full columns, 8-day time-and-strike movement, Lewis patents.
Height 171/2 in. $400
Calendar Clock, E. Ingraham, Bristol, Connecticut, c. 1885. Mahogany veneered case, with carved splat and front, full miniature columns, 8-day time-and-strike movement.
Height 22 in. $550
Calendar Clock, E. Ingraham & Co., Bristol, Connecticut, c. 1885. Pressed oak case, elaborate side frames and splat, gold decorated tablet, figured pendulum, with barometer and temperature gauge to each side, 8-day time-and-strike movement
Height 23 in. $575
American Clock & Watch Museum

Top Left:
Three shelf clocks, from left to right:
Papier-m>chJ Shelf Clock, Forestville Manufacturing Co., Bristol, Connecticut, c. 1855. Black painted case front with excellent gold decoration and inlaid mother-of-pearl, paper label of Daniel Pratt, Reading, Massachusetts, the dealer who originally sold the clock, painted zinc dial, 8-day time-and-strike movement inscribed by the maker.
Height 14 1/2 in. $375
Shelf Clock, unsigned, but probably by Chauncey Jerome, New Haven, Connecticut, c. 1855. Painted black with mother-of-pearl inlay and gold decorations, painted zinc dial, unusually narrow 8-day time-and-strike movement reached through a door in the back.
Height 15 3/4 in. $195
Shelf Clock, J. C. Brown, Bristol, Connecticut, c. 1855. Black painted wood case decorated with fine gold floral, designs, painted zinc dial with maker's name and location, blue paper label, 30-hour time-and-strike movement.
Height 17 in. $225
Richard A. Bourne Co., Inc.

Center Right:
Three varieties of shelf clock, from left to right:
Rare Fusee Shelf Clock, Chauncey Jerome, New Haven, Connecticut, c. 1855. Wood case painted black, with gold designs and mother-of-pearl inlay, painted zinc dial inscribed by the maker with the location, rare 8-day time-and-strike fusee movement marked by the maker.
Height 13 in. $395
Papier-m>chJ Shelf Clock, Chauncey Jerome, New Haven, Connecticut, c. 1855. Black painted case with mother-of-pearl and painted gold floral decorations, painted zinc dial inscribed "C. Jerome," small maker's label, 8-day time-and-strike movement reached through a small door at the back.
Height 16 3/4 in. $275
Papier-m>chJ Shelf Clock, J. C. Brown, Bristol, Connecticut, c. 1850, case - Litchfield Manufacturing Co., Litchfield, Connecticut, pat. 1850. Black papier-m>chJ case has mother-of-pearl inlay and gold decorations, painted zinc dial inscribed "J. C. Brown, Bristol, Ct., U. S.," a small painting of J. C. Brown in the lower bezel, 8-day time-and-strike movement and two paper labels, one for the movement and one for the case.
Height 14 1/2 in. $225
Richard A. Bourne Co., Inc.

On the left is a fine example of a Rosewood Double Dial Mantel Clock, Ithaca Calendar Clock Company, Ithaca, New York, c. 1870. This is a No. 5, with Round Top, housing a brass 8-day spring driven time-and-strike movement with H. B. Horton's calendar mechanism. The calendar rolls have been replaced.
Height 22 1/2 in. $795
The clock on the right is a Walnut Double Dial Shelf Clock, Ithaca Calendar Clock Company, Ithaca, New York, c. 1866. A "No. 9 Shelf Cottage," housing a brass 8-day time-and-strike movement and H. B. Horton's calendar mechanism, with upside-down pendulum aperture. The time dial has been replaced.
Height 22 1/4 in. $950
Robert W. Skinner, Inc.

Three fine calendar shelf clocks, from left to right:
Calendar Clock, Ithaca Clock Co., Ithaca, N. Y. Pat. 1866. "Shelf Steeple" in a walnut case, two paper on zinc dials, the lower with day and month apertures, and date indicator; separate calendar mechanism in lower section of case. "Pat'd Aug 1866" on lower dial, lower dial and movement signed by makers, 8-day time, strike-and-alarm movement by E. N. Welch, Bristol, Connecticut.
Height 25 1/2 in. $1150
Calendar Clock, Ithaca Clock Co., Ithaca, N. Y., c. 1870. "No 10 Farmers" Model, walnut case, paper on zinc dial, the lower with date indicator and day and month apertures, scarce 30-day double spring time movement.
Height 21 in. $775
Calendar Clock, Ithaca Clock Co., Ithaca, N. Y., c. 1870. "No. 5 Round Top Model," lower dial with date indicator and day and month apertures, 8-day time-and-strike movement by N. Pomeroy, Bristol, Connecticut, and so marked.
Height 22 1/2 in. $600
Richard A. Bourne Co., Inc.

Three lovely calendar clocks, from left to right:
Solid Walnut Fashion Model #4 Calendar Clock, Seth Thomas Clock Co., Thomaston, Connecticut, c. 1875. Painted zinc dials, three turned finials, door tablet with gold inscription "Fashion," paper label, 8-day time-and-strike movement with pendulum.
Height 31 1/2 in. $1450
Calendar Clock, Seth Thomas Clock Co., Plymouth Hollow, Connecticut, c. 1865. Early "Column & Cornice" double dial with exceptional rosewood veneered case, columns with gilded capitals, painted zinc dials, instruction label on reverse of the mirror tablet, 8-day time-and-strike movement with iron weights and pendulum, calendar mechanism indicating date, day and month.
Height 32 1/2 in. $1200
Rosewood Veneered Fashion Model #1 Calendar Clock, Seth Thomas Clock Co., Thomaston, Connecticut, c. 1875. Paper on zinc dials, original set-up instructions on inside of door, lower dial inscribed "Southern Calendar Clock Co., St. Louis, Mo.," 8-day time-and-strike movement with pendulum.
Height 28 1/2 in. $1150
Richard A. Bourne Co., Inc.

Three different desirable calendar clocks, from left to right:
"Italian Type" Calendar Clock, with B. B. Lewis Calendar mechanism, Welch Manufacturing Co., Forestville, Connecticut, c. 1870. Rosewood veneered case with beautifully turned columns; 8-day time-and-strike movement. The hour dial has a day indicator, and the lower dial has date and month indicator with a "V" type B. B. Lewis calendar mechanism mounted to the back.
Height 19 3/4 in. $550
"Office Calendar #3" Calendar Clock, Seth Thomas Clock Co., Thomaston, Connecticut, c. 1870. Rosewood veneered case in exceptional condition with painted zinc dials, paper label, 8-day time-and-strike lyre weight driven movement, and a calendar mechanism and bottom section.
Height 33 in. $700
"Shelf Steeple" Calendar Clock, Ithaca Calendar Clock Co., Ithaca, New York, 1870. Solid walnut case with a turned finial and pierced fret surrounding the dials, paper on zinc dials, the lower inscribed by the maker and "Patented Aug. 28, 1866. " 8-day time-and-strike movement, calendar mechanism with date and month apertures and date indicator. Movement by E. N. Welch.
Height 24 1/4 in. $725
Richard A. Bourne Co., Inc.

A desirable Walnut and Ebonized Wood Double Dial Mantel Clock, Ithaca Calendar Clock Co., Ithaca, New York. This is the No. 31/2 Parlor, 2nd model.
Height 20 1/2 in. $2450
Robert W. Skinner, Inc.

Three Calendar clocks, from left to right:
Scarce Double Dial Calendar Kitchen Clock, Waterbury Clock Co., Waterbury, Connecticut, c. 1895. Walnut case with unusual calendar mechanism indicating date and month on the lower dial and day of the week through an aperture in the bottom of the case, 8-day time-and-strike movement with pendulum, painted zinc dials, gold decorated glass.
Height 22 in. $450
Victorian Calendar Clock, E. N. Welch, Forestville, Connecticut, c. 1883. Featuring Franklin-Morse Patent Perpetual calendar mechanism indicating date by a pointer and month on an engraved brass ring through an aperture in the dial. Exceptional carved walnut case with a gold decorated tablet, 8-day time-and-strike movement with pendulum. Two paper labels on back of case.
Height 22 1/2 in. $500
Scarce Calendar Kitchen Clock, William L. Gilbert Clock Co., Winsted, Connecticut, c. 1880. Carved oak case with gold decorated tablet, remnants of a paper label on the back board and a rare calendar dial with a date ring, moon phase dial, and a three pointer hand that moves daily. Day of the month is indicated by the crescent pointer while the small pointers point to the month. Disc must be adjusted manually, 8-day time-and-strike movement with pendulum.
Height 22 in. $525
Richard A. Bourne Co., Inc.

Three different shelf clocks, from left to right:
Rare "Venetian" Shelf Clock, E. Ingraham Clock Co., Bristol, Connecticut, c. 1860. Rosewood veneered case with turned split columns, excellent painted tablet inscribed "E. Pluribus Unum," with an eagle and American shield. Joseph Ives 8-day time-and-strike "Tin Plate" movement with "squirrel cage" roller escapement and rolling pinions.
Height 18 in. $2500
Round Top Shelf Clock, Seth Thomas Clock Co., Thomaston, Connecticut, c. 1875. Rosewood veneered case, painted zinc dial, paper label, maker's signature, 8-day time-and-strike movement.
Height 15 in. $350
Shelf Clock, Jerome & Co., New Haven, Connecticut, c. 1870. Rosewood veneered case, door with gutta percha pressed and gilded design, painted zinc dial, blue paper maker's label and an 8-day time, strike-and-alarm movement.
Height 14 3/4 in. $425
Richard A. Bourne Co., Inc.

Three styles of shelf clocks, from left to right:
Calendar Clock, Burwell & Carter, Bristol, Connecticut, c. 1860. "Venetian" style case of burl walnut with a fine old finish and two dials, the lower indicating the date and the month, and the upper indicating the time and day. 8-day time-and-strike movement and a B. B. Lewis calendar mechanism.
Height 17 in. $475
"Italian #2" Shelf clock, Welch Spring & Co., Forestville, Connecticut, c. 1875. Rosewood veneered case decorated with gilded columns and an excellent black and gold glass, painted zinc dial, paper label, 8-day time-and-strike movement, with "club tooth" escape wheel, and is marked "E. N. Welch & Co., patented 1870. "

Height 17 1/2 in. $350
"Italian #3" Model Shelf Clock, Welch Spring & Co., Forestville, Connecticut, c. 1875. Rosewood veneered case, paper label, painted zinc dial, gold and black floral tablet, 30-hour time movement.
Height 13 1/2 in. $550
Richard A. Bourne Co., Inc.

62.4.110

Venetian Style Round Top Shelf Clock, Welch Spring Co., Forestville, Connecticut, c. 1875. Tablet has decal of bird on cherry tree branch, 8-day brass movement.
Height 17 1/2 in. $450
Museum of Connecticut History

Banjo Shelf Clock, probably Eli Terry & Sons, or Eli Terry Jr., Terryville, Connecticut, c. 1830. One of 10 Connecticut Banjo Clocks known, standard Connecticut banjo case, center tablet of house and landscape, lower tablet mirror, painted face, 30-hour wooden movement.
Height 38 1/2 in. $2850
Museum of Connecticut History

Round Top Shelf Clock, Seth Thomas, Thomaston, Connecticut, c. 1875. Round top rosewood case, mirror, "S" "T" hands, 8-day brass time-and-alarm movement.
Height 15 in. $250
Museum of Connecticut History

62.4.23

Venetian Style Shelf Clock, N. L. Brewster, Bristol, Connecticut, c. 1861. Original dial and tablet, showing the British coat-of-arms in order to appeal to the Canadian market, 8-day rolling pinion movement with rolling escapement, designed and patented.
Height 17 in. $525
Museum of Connecticut History

Venetian Style Round Top Shelf Clock, Welch Spring Co., Forestville, Connecticut, c. 1869-1884. Small round top case, tablet shows gilt on black British seal, 30-hour brass movement with alarm.
Height 14 in. $350
Museum of Connecticut History

62.4.100

62.4.111

Four examples of gingerbread! From the left, they are:

An Oak and Walnut Gingerbread Mantel Clock, Ansonia Clock Company, c. 1880. This the "Jasmine" model, gilt transfer tablet with Americana motif, pseudo-mercury pendulum, 8-day brass spring driven time-and-strike movement.
Height 22 3/8 in. $200

The second clock is a Walnut Gingerbread Mantel Clock, E. N. Welch Manufacturing Company, Forestville, Connecticut, c. 1875. Painted metal dial, gilt transfer with knight and armor motif, 8-day brass spring driven time, strike-and-alarm movement.

Height 23 1/4 in. $295

Third is a Pressed Oak Gingerbread Mantel Clock, Waterbury Clock Company, Waterbury, Connecticut, c. 1880. Applied paper dial, gilt transfer tablet, 8-day brass spring driven time, strike-and-alarm movement.
Height 21 3/4 in. $225

Last is a Walnut Gingerbread Mantel Clock, Ansonia Clock Company, c. 1882. Applied paper dial, 8-day brass spring driven time-and-strike movement, fancy brass calibrated pendulum.
Height 20 3/8 in. $350
Robert W. Skinner, Inc.

Three fine kitchen-style shelf clocks, from left to right:

"Hampden Model" Kitchen Clock, Ansonia Clock Co., Ansonia, Connecticut, c. 1875. Oak case with carved and pressed decorations, silver decorated glass, paper on zinc dial, paper label, 8-day time, strike-and-alarm movement.
Height 21 3/4 in. $225

"Oswego Model" Calendar Clock, Waterbury Clock Co., Waterbury, Connecticut, c. 1890. Solid oak case with pressed designs, paper on zinc dials, complete label and set-up instructions, 8-day time-and-strike movement with date, day and month indicators.
Height 27 3/4 in. $600

"Parisian" Model Kitchen Clock, Ansonia Clock Co., Ansonia, Connecticut, c. 1885. Walnut case with beautifully turned finials, paper on zinc dial, exceptional glass tablet depicting a gentleman with his sword, 8-day time-and-strike spring movement and an unusual pendulum with a figure of a girl applied to an eight-pointed star.
Height 23 1/2 in. $450
Richard A. Bourne Co., Inc.

Three collectible shelf clocks, from left to right:
Kitchen Clock, F. Kroeber, New York, c. 1875. Walnut case with a painted tablet, paper dial, 8-day time-and-strike movement, and a fine pendulum with regulating gauge; pendulum, gong and movement marked "F. Kroeber." Height 18 in. $375

Scarce Calendar Clock, movement probably by Waterbury Clock Co., calendar mechanism by C. W. Feishtinger, Fritztown, Pennsylvania, c. 1894. Walnut case with unusual calendar mechanism patented in 1894 by Charles W. Feishtinger. Painted zinc dials, silver design on the glass tablet; 8-day time-and-strike movement. The lower dial has month and date indicators, and the day is indicated in the aperture in the base molding.
Height 22 in. $900

Kitchen Clock, F. Kroeber, New York, c. 1870. Carved solid walnut case, paper on zinc dial, paper label, 8-day time, strike-and-alarm movement with a fine brass and beveled glass pendulum with regulating gauge.
Height 18 in. $325
Richard A. Bourne Co., Inc.

From the left, the clocks are as follows:
"Admiral George Dewey" Commemorative Shelf Clock, E. Ingraham & Co., Bristol, Connecticut, c. 1899. Flags, cannonballs, stars and anchors are pressed into the oak case, original gold tablet depicting Dewey's ship and flags, gilded lead pendulum with American eagle and shield, 8-day time-and-strike movement with steel plates.
Height 23 in. $325

In the middle is a scarce Oversized Calendar Kitchen Clock, unsigned, but Connecticut, 1880. Carved walnut case with a fine decorated gold tablet, paper-on-zinc dial, 8-day time-and-strike movement with simple calendar mechanism and pendulum.
Height 27 in. $325

On the right is a nice Victorian Shelf Clock, E. N. Welch, Forestville, Connecticut, c. 1880. It features an unusual pendulum embossed with "T & D Lundy, 7 & 9 Third St., San Francisco," fine carved walnut case with three turned finials, paper-on-zinc dial and 8-day time-and-strike movement stamped by Welch.
Height 26 in. $400
Richard A. Bourne Co., Inc.

Two lovely shelf clocks, from left to right:
Scarce Calendar Clock, #31/2 Parlor Model, Ithaca Clock Co., Ithaca, New York, c. 1875. Beautiful walnut case with carved decoration, glass calendar dial, cut glass pendulum bob, 8-day time-and-strike movement inscribed by E. N. Welch and Ithaca Clock Co. Although no examples of this model with a white paper time dial are known to the author, it appears that this dial is original to the clock.
Height 20 1/4 in. $2800

Scarce Shelf Clock, "Cary" Model, Welch, Spring & Co., c. 1880. Beautifully constructed rosewood case with glass sides, paper on zinc dial, maker's label on the back, and an 8-day time-and-strike "Patti" movement with a fancy glass pendulum.
Height 18 in. $950
Richard A. Bourne Co., Inc.

Six mission oak shelf clocks, from left to right:
"Aztec Model" Mission Oak Clock, Seth Thomas Clock Co., Thomaston, Connecticut, c. 1890. Solid oak case with unusual copper dial with white numbers, bird's-eye maple panels around the dial, paper label, 8-day time-and-strike movement with copper pendulum visible below the copper dial.
Height 16 1/2 in. $300
"Sam Martin" Model Mission Oak Shelf Clock, New Haven Clock Co., New Haven, Connecticut, c. 1890. Solid oak case, paper label, applied brass numerals, 8-day time-and-strike movement.
Height 20 1/2 in. $150
Mission Oak Clock, unsigned, Connecticut, c. 1890. Solid oak case with applied brass numerals, 8-day time-and-strike movement.
Height 20 1/2 in. $150
Mission Oak Clock, unsigned, Connecticut, c. 1890. Solid oak case with applied brass numerals and an 8-day time-and-strike movement.
Height 19 in. $125
Mission Oak Clock, New Haven Clock Co., New Haven, Connecticut, c. 1890. Solid oak case, applied brass numerals, paper label, 8-day time-and-strike movement.
Height 13 1/2 in. $125
Mission Oak Clock, Sessions Clock Co., Bristol, Connecticut, c. 1890. Solid oak case, applied brass numerals, 8-day time-and-strike movement.
Height 17 1/2 in. $125
Richard A. Bourne Co., Inc.

Advertising Clock, Baird Clock Co., Plattsburg, New York, c. 1875. "The Blade, Keen Sharp, Toledo, Ohio. " Case painted pale red with white letters, paper on zinc dial inscribed by the maker, 8-day time movement.
Height 30 1/4 in. $2850
Richard A. Bourne Co., Inc.

Three novelty clocks, from left to right:
"Old Mr. Boston" Advertising Clock, Gilbert Clock Co., Winsted, Connecticut, c. 1900. Large sheet iron replicas of a "Mr. Boston" bottle, with an 8-day balance wheel movement. Maker's signature on the dial. Height 21 1/2 in. $450
"Old Mr. Boston" Advertising Clock, Gilbert Clock Co., Winsted, Connecticut, c. 1900. Painted wooden bottle inscribed "Mr. Boston/Fine Liquors. " 8-day balance wheel movement with a silvered dial. Height 9 1/2 in. $450
Rare One-Handed Time piece, One Hand Clock Co., Warren, Pennsylvania, c. 1890. Unusual one-handed time piece with a 30-hour time movement, complete with iron stand. Maker's signature on dial. Height 16 in. $300
Richard A. Bourne Co., Inc.

Two rare wall clocks, from left to right:
Rare Advertising Clock, New Haven Clock Co., New Haven, Connecticut, c. 1900. "Goulding's Manures," paper on zinc dial with maker's name, label on back of case, 8-day time movement.
Height 32 in. $1850
Rare Figure "8" Wall Clock, Seth Thomas Clock Co., Thomaston, Connecticut, c. 1890. Beautifully constructed ash case, black and gold tablet, maker's label, 8-day time movement.
Height 30 in. $1000
Richard A. Bourne Co., Inc.

Two blinking eye clocks, from left to right:
Blinking Eye Clock, unsigned, Connecticut, c. 1860. "Topsy" Model, made in Connecticut. The eyes move in conjunction with the escapement of the movement; 30-hour time movement.
Height 16 1/2 in. $2100
Blinking Eye Clock, unsigned, Connecticut, c. 1860. "Sambo" Model, made in Connecticut. The eyes move in conjunction with the escapement of the movement; 30-hour time balance wheel movement
Height 17 in. $1200
Richard A. Bourne Co., Inc.

Six unusual novelty type clocks, from left to right:
Very rare and unusual Desk Clock, Ansonia Clock Co., Ansonia, Connecticut, New York pat., 1878. A most unusual and interesting little clock with the case front of white metal in the shape of the front of a train including the engineers, an amber reflector, the numerals "45," and a 30-hour balance wheel movement with a paper dial. Maker's signature on dial.
Height 8 in. $900
Rare Novelty Musical Clock, F. Kroeber, New York, c. 1890. A rare and quite possibly unique clock in a red velvet-covered case with brass ornamentation and a ceramic circus Elephant on top. The small crank on the side operates a cylinder musical mechanism on the inside of the box. 30-hour balance wheel movement with a paper dial inscribed by the maker.
Height 7 1/4 in. $550
"Time Flies" Clock, Seth Thomas Clock Co., Thomaston, Connecticut, c. 1900. Nickel plated pressed brass front including floral designs and the inscription "Time Flies," maker's signature on dial.
Height 8 in. $400
Glass Ball Paperweight Clock, New Haven Clock Co., New Haven, Connecticut, c. 1890. Glass ball paperweight clock on a stand; the clock has a brass rim with inset green and clear cut glass jewels, porcelain dial with blue figures and seconds bit. 30-hour time movement. The stand has a green marble base and supports a cast white metal gilded eagle.
Height (of stand) 8 3/4 in. $450
Rare and unusual "Warbler" Novelty Clock, Ansonia Clock Co., Ansonia, Connecticut, c. 1878. Rare novelty clock with a white metal figure of a bird on a branch; on top of the movement case is a bee. The bird retains the original paint. 30-hour clock movement.
Height 4 in. $550
Rare "Lodge" Novelty Alarm Clock, Seth Thomas Clock Co., Thomaston, Connecticut, c. 1890. Rare "Lodge" Model novelty alarm clock in a nickel plated case with nickel and brass decoration. Excellent detail in case including the brick chimney, shingled roof and clapboard siding, paper dial. 30-hour time-and-alarm movement.
Height 7 in. $500
Richard A. Bourne Co., Inc.

Three novelty clocks, from left to right:
Blinking Eye Clock, Bradley & Hubbard, Meriden, Connecticut, c. 1860. "Continental Soldier" Model; the eyes move in conjunction with the escapement of the movement. 30-hour time balance wheel movement, signature stamped into the base.
Height 16 in. $1000
Blinking Eye Clock, T. Kennedy, Connecticut, c. 1860. "John Bull" Model. The eyes move in conjunction with the escapement of the movement; 30-hour time-only balance wheel movement, signature stamped into the base "T. Kennedy, Patent applied for 1856. "
Height 16 1/2 in. $1450
Carved Wooden Eagle Clock, unsigned, c. 1890. Beautifully carved eagle with porcelain dial; 30-hour movement mounted in the center.
Height 15 1/4 in. $500
Richard A. Bourne Co., Inc.

Six novelty clocks, from left to right:
Rare Animated Alarm Clock, Lux Clock Manufacturing Co., Waterbury, Connecticut, c. 1930. Dial with a scene of a man and woman in their home, the woman working on a spinning wheel; the wheel turns in conjunction with the escapement of the movement. Included in the scene is a fireplace and over it is the alarm indicator in the shape of a clock. The dial is colorful and includes a dog in front of the fireplace. Black painted tin case, 30-hour time-and-alarm movement.
Height 4 1/2 in. $150
Rare "Roy Rogers" Animated Alarm Clock, Ingraham Clock Co., Bristol, Connecticut, c. 1940. Retaining the original box, 40-hour movement, painted dial with a scene and Roy on his horse moving back and forth in conjunction with the escapement of the movement.
Height 4 1/2 in. $450
Animated Clock, unsigned, c. 1940. Animated clock with a scene of a windmill on the dial; the wheel of the windmill revolves with the balance wheel of the movement. 30-hour time movement.
Height 4 3/4 in. $100
Animated Desk Clock, Lux Clock Manufacturing Co., Waterbury, Connecticut, c. 1930. Painted scene of a church on the dial, with a bell in the tower that moves in conjunction with the escapement of the 30-hour movement.
Height 4 in. $75
8-Day Alarm Clock, sold by Tiffany & Co., c. 1900. Quality clock with a superb blue decorated enameled dial center, silvered time ring, exceptional hands and a jeweled 8-day movement with an alarm that sounds on a gong inside the case.
Height 4 in. $100
Alarm Clock, sold by Tiffany & Co., c. 1940. Brass case, silvered dial marked "Tiffany & Co.," 8-day 15 jewel movement.
Height 3 in. $75
Richard A. Bourne Co., Inc.

Very Rare Clock Under Dome, F. Kroeber Clock Co., New York, c. 1878. With a "noiseless rotary pendulum. " Unusual escapement with a worm gear, 8-day time-and-strike movement that strikes on a gong mounted under the base. The pendulum is a nickel plated brass ball suspended by a thread.
Height (incl. Dome) 20 in. $2650
Richard A. Bourne Co., Inc.

Two very rare and unusual clocks, from left to right:
Very Rare Domed Candlestand Clock, Terryville Manufacturing Co., Terryville, Connecticut, c. 1854. Produced only for a period of two years, this appealing clock has a milk glass base, original glass dome, 30-hour spring movement with torsion bar control, and a silvered pressed brass dial, inscribed "Ansonia Clock Co., Ansonia, Ct. U. S. A. " which is original to the clock. It is probable that Ansonia Clock Co. bought out Terry's parts after his business failed and then completed this clock under their name. Back cover of movement is also marked by Terryville Manufacturing Co.
Height (including dome) 10 1/4 in. $1500
Very Rare American Skeleton Clock, Terry Clock Co., Waterbury, Connecticut, c. 1875. "Parlour" Model with a glass dome, porcelain dial, pressed brass decoration inscribed by the maker, 8-day double wind time movement. An exceptional feature of this clock is the iron base with the original painted decorations; it also retains the original pendulum and period glass dome.
Height (excluding dome) 9 in. $850
Richard A. Bourne Co., Inc.

Very Rare Domed Candlestand Clock, Terryville Manufacturing Co., Terryville, Connecticut, c. 1852. 30-hour time-only balance wheel movement, with a fine original paper dial and brass spring cover embossed, "Terryville Manufacturing Co., Terryville, Conn., Patented Oct. 5th, 1852. " All on a milk glass base with a glass dome.
Height 10 3/4 in. $3100
Richard A. Bourne Co., Inc.

Three lovely candlestand clocks under domes, from left to right:
Candlestand Clock Under Dome, Seth Thomas Clock Co., New York City, c. 1890. Under dome, with a painted and gilded stand, porcelain dial, 8-day time movement with maker's inscription on the movement.
Height 10 3/4 in. $900
Candlestand Clock Under Dome, unsigned, c. 1880. 30-hour time movement with a fixed pendulum mounted on a porcelain stand and turned wooden base.
Height 9 1/2 in. $400
Candlestand Clock Under Dome, Seth Thomas Sons & Co., New York City, 1880. With a turned, painted and gilded support, porcelain dial with a brass bezel, 8-day time-and-strike movement with maker's signature on movement.
Height 10 in. $750
Richard A. Bourne Co., Inc.

Very Rare Candlestand Clock, Terryville Manufacturing Co., Terryville, Connecticut, c. 1855. 30-hour time-only movement, with Terry's patented torsion balance movement, a brass plate covering the main spring with maker's name and patent date of "Oct. 5th, 1852" embossed on it, all on a turned walnut base with a glass dome.
Height 10 3/4 in. $1700
Richard A. Bourne Co., Inc.

Here we have four fine examples of shelf clocks, the first on the left being a Mahogany Veneer Triple Decker Shelf Clock, c. 1832. It bears a label reading "Abraham & Albert, Landisburg, Pa. " It has a cornice top, gilt highlighted painted wooden dial, mirror tablet over reverse painted tablet, freestanding columns over ogee plinth, the whole raised on turned feet, housing a brass strap 8-day time-and-strike movement The piece has had restoration.
Height 36 in. $700
Second from the left is a Rosewood Veneer Shelf Clock, Seth Thomas, Thomaston, Connecticut, c. 1880. The case is similar to the Parlor Calendar No. 1, reverse painted door tablet, housing a brass lyre 8-day weight driven movement. There has been restoration.
Height 35 in. $650
The third clock is a Carved Mahogany Triple Decker Shelf Clock, C. & L. C. Ives, Bristol, Connecticut, c. 1840. There is a cornucopia carved crest, gilt highlighted painted wooden dial, mirror center tablet flanked by short carved columns over a black and gold reverse painted tablet, the whole raised on ball feet, housing a brass strap 8-day time-and-strike movement. Restoration has been done.
Height 38 in. $600
The clock on the right is a Mahogany Veneer Triple Decker Shelf Clock, Birge & Fuller, Bristol, Connecticut, c. 1843. It has a gilt crest, floral decorated painted wooden dial, grain-painted half columns with gilt capitals and bases, geometric reverse painted center tablet, mirrored lower door, housing a brass weight driven time-and-strike movement. The finials have been replaced and the feet are missing.
Height 34 1/4 in. $650
Robert W. Skinner Inc.

From left to right:
A Mahogany Cased Looking Glass Shelf Clock, Daniel Pratt Jr., Reading, Massachusetts, c. 1830. Painted wooden dial, flat pilasters, housing a 30-hour weight driven movement. Minor veneer damage.
Height 34 1/2 in. $225
The next clock is a Mahogany Veneer Looking Glass Shelf Clock, M. & E. Blakeslee, Heathersville, Connecticut, c. 1830. Shaped crest over painted wooden dial and mirror, flanked by stenciled baluster shaped columns, housing a 30-hour weight driven movement.
Height 32 in. $225
The third clock is a Mahogany Veneer Ogee Mantel Clock, Chauncey Goodrich, Bristol, Connecticut, c. 1830. It has a painted zinc dial, poly-chrome transfer door tablet "Public Square New Haven", housing a brass 30-hour weight driven time, strike-and-alarm movement. Partial label.
Height 29 in. $325
The last clock on the right is a Mahogany Veneer Looking Glass Shelf Clock, Pratt & Frost, Reading, Massachusetts, c. 1832. It has a stenciled crest over the painted wooden dial and mirror tablet, flanked by gilt banded half columns, using an 8-day weight driven wooden movement.
Height 34 3/4 in. $325
Robert W. Skinner Inc.

On the left is a Mahogany Looking Glass Clock, Chauncey Boardman, Bristol, Connecticut, c. 1830. It has a stenciled crest, painted and deco-rated wooden dial, stenciled baluster shaped columns, housing a 30-hour weight driven movement.
Height 34 3/4 in. $400
The middle clock is a Mahogany Veneer Looking Glass Shelf Clock, David Dutton, Mount Vernon, New Hampshire, c. 1825. It has a stenciled crest, painted and decorated wooden dial, stenciled half columns, housing a 30-hour weight driven time-and-strike wooden movement. Minor veneer dam-age.
Height 33 1/4 in. $250
The clock on the right is an Empire Mahogany Veneer Triple Decker Shelf Clock, Birge, Mallory & Co., Bristol, Connecticut, c. 1840. There is a cornice top, gilt decorated wooden dial, free standing columns over curved pilas-ter, reverse painted lower door tablet, housing a brass 8-day time-and-strike strap movement. The upper door tablet is missing and there is other damage.
Height 35 in. $450
Robert W. Skinner Inc.

From left to right:
Ingraham Kitchen Clock, E. Ingraham Co., Bristol, Connecticut, first quar-ter twentieth century. Pressed oak case, 8-day time-and-strike movement.
Height 23 in. $200
In the middle is a fine Sleigh Front Calendar Clock, Seth Thomas, Thomaston, Connecticut, mid-nineteenth century. Rosewood facings with partially gilded columns, 8-day weight driven time-and-strike movement.
Height 32 3/8 in. $600
Shelf Clock, Jerome & Darrow, Bristol, Connecticut, c. 1830. Transitional case with stenciled half columns and splat, carved pineapple finial and paw feet.
Height 30 in. $600
Richard A. Bourne Co., Inc.

Three shelf clocks, from left to right:
Empire Shelf Clock, Birge & Mallory, Bristol, Connecticut, c. 1840. Mahogany veneered case with ball feet, painted tablet, maker's label, fine large 8-day strap brass weight driven movement stamped "B. M. & Co.," with roller pinions; opening in dial allows for adjustment of escapement.
Height 37 in. $450
Split Column Shelf Clock, Seth Thomas Clock Co., Plymouth, Connecticut, c. 1830. Mahogany case with beautifully stenciled columns and an eagle splat, paper label inscribed by the maker, painted wood dial, 30-hour time-and-strike wood movement and a mirror in the door.
Height 34 in. $500
Split Column Shelf Clock, Seth Thomas Clock Co., Plymouth Hollow, Connecticut, c. 1830. Mahogany veneered case with an excellent original tablet, colorful painted wood dial, maker's label, 30-hour time-and-strike wood movement.
Height 34 3/4 in. $350
Richard A. Bourne Co., Inc.

Three fine shelf clocks, from left to right:
Split Column Shelf Clock, Edward Barnes, Bristol, Connecticut, c. 1830. Mahogany case with stenciled columns and splat, Adams print in the door, painted wood dial, paper label, 30-hour time-and-strike wood movement with weights.
Height 31 3/4 in. $325
Empire Shelf Clock, L. S. Hotchkiss & Co., New York, c. 1840. Mahogany veneered case with split columns, mirror, painted wood dial, 8-day time-and-strike Forestville-type movement.
Height 30 3/4 in. $350
Split Column Shelf clock, Pratt & Frost, Reading, Massachusetts, c. 1835. Mahogany case with mirror, painted wood dial, paper label and 30-hour time-and-strike wood "Groaner" movement.
Height 35 in. $325
Richard A. Bourne Co., Inc.

Three fine shelf clocks, from left to right:
Shelf Clock, Spencer Hotchkiss & Co., Salem Bridge, Connecticut, c. 1825. Choice mahogany veneered case with superb painted wood dial with seconds indicator, paper label, 8-day time-and-strike "Salem Bridge" weight driven movement.
Height 33 in. $750
Pillar and Scroll Clock, Eli Terry & Sons, Plymouth, Connecticut, c. 1830. Mahogany case with turned columns, scrolled crest and feet, painted tablet, painted wood dial, paper label, 30-hour time-and-strike wood movement.
Height 31 3/4 in. $1000
Split Column Shelf Clock, Seth Thomas Clock Co., Thomaston, Connecticut, c. 1870. Rosewood veneered case with exceptional paint decorated columns, mirror, painted zinc dial, 30-hour time-and-strike movement with period weights.
Height 25 in. $375
Richard A. Bourne Co., Inc.

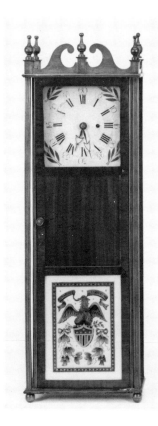

Extremely Rare And Early Shelf Clock, attributed to John Taber, Saco, Maine, c. 1815. Unusual slender design with a maple case and a mahogany veneered door, delicate scrolled crest with turned finials, fine eglomise tablet, exceptional painted wood dial, 8-day time movement with brass bushed iron plates. An interesting feature is that the plates of the movement are held together by the long screws that shoulder on the front plate and attach the movement to the case. The case is stamped "R. B. Taft. "
Height 37 1/2 in. $4200
Richard A. Bourne Co., Inc.

Two rare Silas B. Terry shelf clocks, from left to right:
Rare Empire Style Shelf Clock, Silas B. Terry, Terryville, Connecticut, c. 1835. Fine mahogany case with carved feet, columns and crest; the lower door contains the original painted drape that frames an eglomise panel applied to the backboard. 8-day weight-driven movement with count wheel strike has a solid brass back plate and a pierced brass front plate; retains two iron weights, old pendulum and original crank key.
Height 37 1/2 in. $1250
Shelf Clock, Silas B. Terry, Terryville, Connecticut, c. 1835. Mahogany case with a fine carved crest and carved feet; exceptional wood dial, 8-day time-and-strike weight driven movement with a solid backplate and a pierced front plate. Opening in the dial reveals the movement and an engraved brass seconds indicator.
Height 38 in. $750
Richard A. Bourne Co., Inc.

Two shelf clocks, from left to right:
Rare Shelf Clock, Silas B. Terry (attributed), Terryville, Connecticut, c. 1835. Exceptional carved mahogany basket of fruit on the crest, carved feet, painted tablet of lady with gold leaf border, fine enameled wood dial with gold decorations, 8-day time-and-strike movement with pierced front plate and solid backplate retaining period iron weights and pendulum.
Height 38 1/2 in. $900
Rare Shelf Clock, Hotchkiss & Benedict, Auburn, New York, c. 1830. Mahogany veneered case with carved crest and capitals, painted tablet of a castle, paper label, exceptional black painted dial with gold decorations inscribed by the maker, with seconds indicated by a brass finger, and with a 8-day time-and-strike movement number "2855" with flying eagle pendulum and iron weights.
Height 37 1/2 in. $1000
Richard A. Bourne Co., Inc.

Two lovely shelf clocks, from left to right:
Shelf Clock, Seth Thomas Clock Co., Thomaston, Connecticut, c. 1875. "Garfield" Model of solid oak with brass weights, pendulum and bell mount, paper label, 8-day time-and-strike movement.
Height 29 1/2 in. $950
Shelf Clock, Chauncey Boardman & Joseph Wells, Bristol, Connecticut, c. 1835. Mahogany case with exceptional carved columns and crest, mirror tablet, paper label, painted wood dial and a 30-hour time-and-strike wood movement with iron weights.
Height 34 1/2 in. $700
Richard A. Bourne Co., Inc.

Three rare shelf clocks, from left to right:
Carved Shelf Clock, Seymour, Williams & Porter, Unionville, Connecticut, c. 1835. Fine mahogany case with beautifully carved "basket of fruit" crest, columns, and paw feet. Enameled wood dial, paper label, unusual 8-day time-and-strike wood movement with strike mechanism on the bottom half of the plates and time mechanism on the top half, retains weights and pendulum.
Height 37 1/2 in. $850
Very Rare Triple Decker Wagon Spring Shelf Clock, C. & L. C. Ives, Bristol, Connecticut, c. 1835. An extremely rare model using Joseph Ives Patent lever spring mounted in the bottom of the case. Case is constructed of mahogany and mahogany veneer with ebonized ball feet and columns, decorated eagle crest, paper label, 8-day time-and-strike strap brass movement with pendulum, enameled wood dial.
Height 36 1/2 in. $1200
Rare Shelf Clock, C. & L. C. Ives, Bristol, Connecticut, c. 1835. Featuring an early 8-day time-and-strike movement said to be the first movement used after Ives left New York, having gears similar to those used in the looking glass clocks with lead winding drums. Mahogany case with stenciled columns and splat, paper label, mirror tablet, 2 iron weights and pendulum.
Height 35 in. $950
Richard A. Bourne Co., Inc.

Three shelf clocks, from left to right:
Quarter Column Shelf Clock, Eli Terry & Sons, Terryville, Connecticut, c. 1825. Excellent stenciled columns and splat, 30-hour wood time-and-strike movement.
Height 28 in. $700
Pillar and Scroll Clock, Jerome & Darrow, Bristol, Connecticut, c. 1825. Tablet with landscape and classical building, with aperture, 30-hour wood time-and-strike movement.
Height 29 in. $900
Connecticut Shelf Clock, Edward K. Jones, Bristol, Connecticut, c. 1835. 30-hour wood time-and-strike movement.
Height 27 in. $650
Richard A. Bourne Co., Inc.

Three shelf clocks, from left to right:
Rare Miniature Triple Decker Shelf Clock, Birge, Mallory & Co., Bristol, Connecticut, c. 1835. Excellent original tablets, 30-hour brass time-and-strike weight driven movement.
Height 26 in. $850
Shelf Clock, no maker shown, c. 1830. Superbly carved fruit crest and paw feet, 8-day brass Salem Bridge movement, weight driven with second bit.
Height 30 3/4 in. $500
Shelf Clock, Charles Kirk, Bristol, Connecticut, c. 1833. Eagle crest, 30-hour weight driven wooden movement, stenciled columns, fine original tablet.
Height 28 1/2 in. $550
Richard A. Bourne Co., Inc.

Three different style shelf clocks, from left to right:
Large Shelf Clock, E. Thayer, Williamsburg, Massachusetts, c. 1830. Carved pillars and splat, landscape and house scene in door glass, 30-hour wooden Groaner movement, time-and-strike.
Height 35 in. $450
Connecticut Shelf Clock, Silas Hoadley, Plymouth, Connecticut, c. 1830s. Very fine label, and dial, stenciled columns and splat, 30-hour upside-down wooden time, strike-and-alarm movement.
Height 36 in. $950
Empire Shelf Clock, Seth Thomas, Thomaston, Connecticut, c. 1860. Excellent rosewood case with fine label, 8-day weight driven brass time-and-strike movement with perpetual calendar at the bottom.
Height 33 in. $900
Richard A. Bourne Co., Inc.

Three fine shelf clocks, from left to right:
Shelf Clock, unknown maker, c. 1830. Mahogany case with fine carved columns, 30-hour wooden Terry-type works, time-and-strike.
Height 28 in. $400
Shelf Clock, E. G. & W. Bartholomew, Bristol, Connecticut, c. 1830. Carved pineapple finials and carved paw feet, Terry patent 30-hour wooden time-and-strike movement.
Height 25 1/2 in. $400
Shelf Clock, Silas Hoadley, Plymouth, Connecticut, c. 1835. Excellent dial and label, lower glass door original, but mostly flaked off; 30-hour time-and-strike movement with ivory bushings.
Height 28 in. $625
Richard A. Bourne Co., Inc.

Round Top Shelf Clock, Aaron Willard, Boston, Massachusetts, c. 1800. Mahogany veneered case, painted zinc dial with "Aaron Willard, Boston" in gold leaf surround. 8-day time and strike movement.
Height 25 1/2 in. $1300
American Clock & Watch Museum

"Venetian Model" Round Top Shelf Clock, E. Ingraham & Co., Bristol, Connecticut, c. 1870. Rosewood veneered case, wood lower sash, brass upper sash, attractive gold leaf tablet, paper on zinc dial, 8-day brass time-and-strike spring driven movement.
Height 18 in. $450
American Clock & Watch Museum

Round Top Shelf Clock, Welch, Spring & Co., Bristol, Connecticut, c. 1875. Rosewood veneered case, painted zinc dial, painted tablet, 8-day time, strike-and-alarm brass movement.
Height 18 in. $450
American Clock & Watch Museum

Round Top Shelf clock, Atkins & Son, Bristol, Connecticut, c. 1860. Walnut veneered case, lovely black and gold glass, painted zinc dial, 8-day brass time, strike-and-alarm movement.
Height 15 in. $400
American Clock & Watch Museum

"Venetian" Style Shelf Clock, Atkins Clock Co., Bristol, Connecticut, c. 1875. Mahogany veneered case gold leafed arms tablet, paper dial in brass bezel, 8-day time-and-strike brass movement.
Height 17 in. $400
American Clock & Watch Museum

"Masonic" Model Shelf Clock, E. N. Welch Manufacturing Co., Bristol, Connecticut, c. 1865. Mahogany veneered case showing Masonic emblems, paper on zinc dial, 8-day time-and-strike brass movement.
Height 17 in. $450
American Clock & Watch Museum

"Doric Mosaic" Model Shelf Clock, E. Ingraham & Co., Bristol, Connecticut, c. 1878. Rosewood veneered case, Mosaic front, paper on zinc dial, lovely black and gold tablet with floral motif, 8-day time-and-strike brass movement.
Height 16 1/4 in. $395
American Clock & Watch Museum

"Connecticut Banjo" Shelf Clock, Eli Terry & Sons, Bristol, Connecticut, c. 1820. Mahogany case, painted wood dial, classical building and foliage scene on tablet, 30-hour time-and-strike movement.
Height 35 in. $1500
American Clock & Watch Museum

"Doric" Model Shelf Clock, E. Ingraham & Co., Bristol, Connecticut, c. 1880. Rosewood veneered case, attractive painted tablet, paper on zinc dial, 8-day brass time-and-strike movement.
Height 16 in. $450
American Clock & Watch Museum

Free Standing Banjo Shelf Clock, unsigned, Massachusetts, c. 1845. Mahogany veneered case, painted metal dial, 8-day time-only weight driven movement.
Height 30 in. $1000
American Clock & Watch Museum

"Dauphin" Crystal Regulator Mantel clock, Ansonia Clock Co., New York, c. 1906. White onyx top and base with beveled glass front, back and sides, standing figurines on all four corners, rich gold finish, porcelain dial with open escapement, 8-day movement with mercury pendulum and cathedral gong.
Height 18 1/2 in. $2100
Lindy Larson

Rear view of the "Dauphin" Crystal Regulator Clock.
Lindy Larson

"Orleans" Crystal Regulator Mantel Clock, William L. Gilbert Clock Co., Winsted, Connecticut, c. 1905. A most unusual case style; all four sides and the top are one piece of crystal, the cast metal top and base trim are applied over the crystal and then gold plated. 4 inch double sunk porcelain dial with open escapement, 8-day movement with cathedral gong and mercury pendulum.
Height 11 in. $1000
Lindy Larson

Rear view of the "Orleans" Crystal Regulator Clock.
Lindy Larson

"Fortuna" Swinging Arm Model Shelf Clock, Ansonia Clock Co., New York, c. 1906. This cast metal statuary came with the option of several types of bronze finishes; the clock portion also came in several finishes. The clock pivots on the statue's outstretched hand and forms its own pendulum as it swings back and forth; 4 inch dial with brass hands, 8-day timepiece.
Height 30 in. $3800
Lindy Larson

Timby's Solar Time Piece, L. E. Whiting, Saratoga Springs, New York, c. 1870. Mahogany case, rotating globe with circular dial in lower sash.
Height 28 in. $1100
American Clock & Watch Museum

Shelf Clock, Kroeber Clock Co., New York, c. 1888. Mahogany veneered case, painted metal dial, rotary (noiseless) pendulum clock, 8-day brass time-and-strike movement.
Height 21 in. $600
American Clock & Watch Museum

"Sibyl and Gloria" Statue Clock, Ansonia Clock Co., New York, c. 1900. Cast metal statuary with two tone bronze finish, 4 inch porcelain dial with beveled glass, 8-day pendulum movement with gong strike.
Height 27 1/2 in. $1500
Lindy Larson

"Gloria" Model Shelf Clock, Ansonia Clock Co., Ansonia, Connecticut, c. 1900. Swinging ball or pendulum style clock, 8-day brass time-and-strike movement.
Height 29 in. $2600
American Clock & Watch Museum

Rear view of the "Sibyl and Gloria" Statue Clock.
Lindy Larson

"Eureka" Model Shelf Clock, Ansonia Clock Co., Ansonia, Connecticut, c. 1910. Ornate white metal plated case, heavily ornamented with metal fittings, 8-day brass time-and-strike movement, Brocot type escapement.
Height 22 in. $1250
American Clock & Watch Museum

"Lydia" Clock set, Ansonia Clock Co., New York, c. 1894. Cast metal clock and candelabra, bronze or silver plated, 8-day movement with cathedral gong strike, open escapement.
Height 19 1/2 in. $1600
Lindy Larson

Close-up of the clock portion of the "Lydia" candelabra set.
Lindy Larson

Victorian Style Parlor Clock, Terry Clock Co., Pittsfield, Massachusetts, c. 1880. Rosewood case with carved grapes, leaves and flowers above and below the dial, signed porcelain dial, cast brass bezel with heavy beveled glass, 8-day movement.
Height 16 in. $450
Lindy Larson

Victorian Style Parlor Clock, Ansonia Clock Co., Ansonia, Connecticut, c. 1875. Walnut case with turned finials and applied rosettes, painted zinc dial, stenciled glass in the pattern of curtains showing the very ornate pendulum. The 8-day movement has a patented self leveling suspension in place of the usual suspension spring.
Height 22 in. $450
Lindy Larson

"Monarch" Model Shelf Clock, William L. Gilbert Clock Co., Winsted, Connecticut, c. 1880. Large walnut case with applied carvings and a secret drawer in the base, heavy, ornate cast brass dial pan and fancy cut glass pendulum, stenciled glass, 8-day movement with nickeled bell strike.
Height 24 in. $600
Lindy Larson

Victorian Parlor Clock, George B. Owen, New York, c. 1875. Solid walnut case with double sets of carved side trim pieces, cut glass tablet, paper dial and an ornate brass and nickel pendulum, 8-day movement with bell strike. Height 23 in. $500
Lindy Larson

"Victorian Style" Shelf Clock, Forestville Manufacturing Co., Bristol, Connecticut, c. 1837. Crotch mahogany veneered case, including the scrolled base, molded top and 3/4 columns, three sashes, all with deeply etched and polished plate glass, painted metal dial, 8-day brass time-and-strike movement.
Height 42 in. $1350
American Clock & Watch Museum

"Cabinet #7" Kitchen Clock, E. Ingraham & Co., Bristol, Connecticut, c. 1895. Heavily carved and decorated pressed oak case, decorated bezel, painted zinc dial, 8-day time-and-strike brass movement.
Height 15 1/2 in. $650
American Clock & Watch Museum

"Cayohga" Model kitchen Clock, E. Ingraham & Co., Bristol, Connecticut, c. 1903. Pressed oak case, paper on zinc dial, lovely gold stenciled tablet, 8-day brass time-and-strike movement.
Height 22 1/2 in. $550
American Clock & Watch Museum

Kitchen Clock, E. Ingraham & Co., Bristol, Connecticut, c. 1900. Pressed oak case with thermometer attached to splat, paper on zinc dial, black and gold tablet with "Time Is Money," Benjamin Franklin's favorite expression, 8-day time-and-strike brass movement.
Height 24 in. $450
American Clock & Watch Museum

Kitchen Clock, Sessions Clock Co., Bristol, Connecticut, c. 1900. Pressed oak case, beautifully decorated gold and black floral tablet, paper on zinc dial, 8-day time-and-strike brass movement.
Height 24 in. $500
American Clock & Watch Museum

"Santa Fe" Model Shelf Clock, Seth Thomas Clock Co., Thomaston, Connecticut, c. 1880. One of the rare models in the city series issue by Seth Thomas. Walnut case with carved lattice work, applied metal leaves, matching stenciled glass of grapes, vines and leaves, nickeled and damascened pendulum, painted zinc dial, 8-day time movement with cathedral gong.
Height 22 1/2 in. $595
Lindy Larson

Kitchen Clock, E. N. Welch Manufacturing Co., Bristol, Connecticut, c. 1900. Pressed oak case, ornate black and gold tablet, paper on zinc dial, 8-day time-and-strike brass movement.
Height 24 in. $600
American Clock & Watch Museum

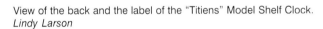

Close-up view of the porcelain dial and pendulum of the "Titiens" Model Shelf Clock.
Lindy Larson

View of the back and the label of the "Titiens" Model Shelf Clock.
Lindy Larson

"Titiens" Model Shelf Clock, Welch, Spring & Co., Forestville, Connecticut, c. 1880. Rosewood case of superior quality, similar in style to the "Regulator #5" with fully turned columns, drop finials and glass sides with gold leaf moldings, black flocked backboard, seven inch porcelain dial signed "Welch, Spring & Co., Forestville, Conn. " in script. 30-day double wind nickel plated movement has a club tooth escapement and is signed. Height 23 in. $2650
Lindy Larson

Kitchen Style Shelf Clock, William L. Gilbert, Winsted, Connecticut, c. 1885. Walnut veneered case with full sash, turned and carved drops, finials and ornaments, the lower part of the sash is silver stenciled, 8-day brass time, strike-and-alarm spring driven with center alarm set movement. The original price is noted on the attached tag.
Height 24 in. $550
American Clock & Watch Museum

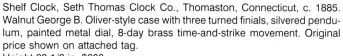

Shelf Clock, Seth Thomas Clock Co., Thomaston, Connecticut, c. 1885. Walnut George B. Oliver-style case with three turned finials, silvered pendulum, painted metal dial, 8-day brass time-and-strike movement. Original price shown on attached tag.
Height 23 1/2 in. $600
American Clock & Watch Museum

"Manhattan" Model Shelf Clock, Ansonia Clock Co., Ansonia, Connecticut, c. 1885. Walnut veneered case, painted metal dial, simulated mercury pendulum, 8-day brass time-and-strike movement. Original price shown on attached tag.
Height 24 in. $550
American Clock & Watch Museum

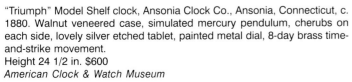

"Triumph" Model Shelf clock, Ansonia Clock Co., Ansonia, Connecticut, c. 1880. Walnut veneered case, simulated mercury pendulum, cherubs on each side, lovely silver etched tablet, painted metal dial, 8-day brass time-and-strike movement.
Height 24 1/2 in. $600
American Clock & Watch Museum

"Parisian" Model Scrolled Kitchen Shelf Clock, E. N. Welch, Bristol, Connecticut, c. 1880. Walnut veneered case, carved, turned and scrolled, stenciled sash, ornamental bob, 5 inch painted metal dial, 8-day time. strike-and-alarm brass movement.
Height 24 in. $450
American Clock & Watch Museum

"Granville" Model Shelf Clock, Waterbury Clock Co., Waterbury, Connecticut, c. 1875. Walnut case in an unusual design with raised moldings, finials and sides set a 45 degree angle, stenciled glass, painted zinc dial and a nickeled pendulum, 8-day movement with cathedral gong.
Height 22 in. $425
Lindy Larson

"President McKinley" Commemorative Shelf Clock, E. Ingraham Co., Bristol, Connecticut, c. 1905. Pressed oak case with embossed cannons, flags, anchors, stars, etc. Bust of President McKinley is a separate applied piece, stenciled patriotic glass of a battleship and flags, special pendulum for this series with an embossed eagle, signed paper dial, 8-day movement.
Height 25 in. $600
Lindy Larson

"Monarch" Style (earlier called "Champion") Shelf Clock, Ansonia Clock Co., Ansonia, Connecticut, c. 1878. Fancy walnut veneered case with drawer in the bottom, applied metal ornaments, fancy glass tablet, 8-day brass time-and-strike movement, Geneva stops.
Height 25 in. $650
American Clock & Watch Museum

"Admiral Sampson" Commemorative Shelf Clock, E. N. Welch Manufacturing Co., Forestville, Connecticut, c. 1905. Pressed oak case with the admiral depicted on the crest, with helmsmen, cannon, etc. below. There were six models in this series by Welch, all of which are quite rare; the others are "Admiral Dewey," Admiral Schley," "Lee," "Wheeler," and the "Battleship Maine. " All models had different case designs and different stenciled glasses. The glass on this particular clock is a replacement of an incorrect design. 8-day movement.
Height 24 in. $700
Lindy Larson

Interior view of the "Admiral Sampson" Commemorative Shelf Clock.
Lindy Larson

"Dewey" Medallion Model Shelf Clock with steel die used in forming splat, E. Ingraham & Co., Bristol, Connecticut, c. 1900. Pressed oak case and splat, black and gold tablet with nautical scene commemorating Dewey's victory at Manila Bay, Dewey splat with patriotic motif, painted metal dial, 8-day time-and-strike brass movement.
Height 23 in. $750
American Clock & Watch Museum

"Admiral Dewey" Commemorative Shelf Clock, E. Ingraham Co., Bristol, Connecticut, c. 1905. Pressed oak case and stenciled battleship glass have the same patriotic symbols as others in this series; the applied bust on the crest depicts Admiral Dewey, a hero of the Spanish American War, eagle decorated pendulum, signed paper dial, 8-day movement.
Height 25 in. $450
Lindy Larson

Close-up view of the crest of the "Emblem" Commemorative Shelf Clock, showing how the emblem is a separately applied pressing of wood.
Lindy Larson

"Emblem" Commemorative Shelf clock, E. Ingraham Co., Bristol, Connecticut, c. 1905. Pressed oak case with patriotic designs; other models in this series which used the same case are: "President McKinley," "Admiral Dewey," "Battleship Maine," "Lee," and "Peace. " Some variations were used in the pendulums and the stenciled glasses. 8-day movement with optional alarm.
Height 25 in. $500
Lindy Larson

Oak Shelf Clock, E. Ingraham Co., Bristol, Connecticut, c. 1905. Commonly called the "Labor and Industry" model; the oak case has an applied piece on the crest with an anvil, scythe and sheaf of wheat. Figures at the sides are shown tending gardens and working at a forge; paper dial, stenciled glass, 8-day brass movement.
Height 25 in. $500
Lindy Larson

"Pirate" Alarm Clock, Ansonia Clock Co., New York, c. 1900. Nickeled case with bell on top, signed paper dial, shown here in its original wood carton. One-day timepiece.
Height 6 in. $200
Lindy Larson

Miniature Shelf Clock, Silas Hoadley, Plymouth, Connecticut, c. 1825. Unusual case style with stenciled splat and quarter columns, painted bird's-eye maple sides and mahogany veneer trim, painted wood dial and mirror below, one-day time-and-alarm wood movement.
Height 23 in. $1750
Lindy Larson

Interior view of the Silas Hoadley Miniature Shelf Clock.
Lindy Larson

Advertising Regulator Clock, New Haven Clock Co., New Haven, Connecticut, c. 1880. Rosewood veneered case, printed paper dial with black on red center section advertising Lucky Strike tobacco. Lower tablet with "Regulator. "
Height 24 in. $700
Lindy Larson

Rear view of the advertising clock. The label indicates that this clock was given as a promotion to the retailer who purchased R. A. Patterson products.
Lindy Larson

Very Rare Parade Clock, Bradley & Hubbard, Meriden, Connecticut, c. 1885. A unique clock with a case of stamped sheet brass, embellished with assorted sizes of colored glass; translucent glass dial reveals time at night by a lighted candle inside of case. Clock was carried in torchlight parades and hung outside for night-time picnics; 30-hour time-only balance wheel movement.
Height 14 in. $1800
Richard A. Bourne Co., Inc.

Advertising Clock, Sidney Advertising Clock Co., c. 1885. Unstained mahogany case with rippled molding, painted metal dial, advertising panel (unfortunately flaking !), lower panel has advertisements that revolve automatically every five minutes; 8-day time-and-strike movement. Highly collectible !
Height 54 in. $7500
American Clock & Watch Museum

Shelf Clock, unsigned, origin unknown, c. 1930s. Marble and bronze dore shelf clock, rectangular clock being held by two oriental figures in flowing robes.
Height 11 in. $400
Winter Associates

Rear view of the Ansonia Clock Co., Swinging Doll Clock.
Lindy Larson

"Swing No. 1 Model" Swinging Doll Clock, Ansonia Clock Co., Ansonia, Connecticut, c. 1890. Alarm clock type base of pressed brass. An arm goes from the escapement through one of the upright posts and pushes the rod on which the cupid is suspended; the hand-painted porcelain cupid swings with each beat of the escapement. One-day timepiece.
Height 12 in. $1150
Lindy Larson

Domed Shelf Clock, Hiram Camp, New Haven, Connecticut, c. 1850. Prototype shelf clock, white marble base, painted dial, 8-day time-and-strike fusee movement.
Height (incl. dome) 21 in. $450
American Clock & Watch Museum

Briggs Rotary Pendulum Shelf Clock, E. N. Welch, Forestville, Connecticut, c. 1878. Black painted wood base with iron feet and the winding wheel underneath, paper dial. The pendulum ball is suspended by a silk thread and moved by a fine spring arm below. One-day movement.
Height (incl. dome) 8 in. $1000
Lindy Larson

"*Dickory, Dickory Dock,*" Dungan & Klump, Philadelphia, Pennsylvania, c. 1910. An interesting novelty clock; the wood case has vertical brass numerals from 1 to 12, as the clock runs, a wooden mouse is carried up the ladder chain through a slot in the front. When the clock strikes 1, the mouse falls down to begin his 12 hour trip again. Two-day lever movement by New Haven.
Height 43 in. $1200
Lindy Larson

"Parlor No. 1" Model Shelf Clock, Seth Thomas Clock Co., Thomaston, Connecticut, c. 1870. Rosewood veneered case, painted metal dial 8-day brass time-and-strike weight driven movement. Absolutely superb condition !
Height 33 in. $1100
American Clock & Watch Museum

Close-up of the "Dickory, Dickory, Dock" timepiece showing the mouse and stenciled nursery rhyme.
Lindy Larson

Rippled Cottage Clock, J. C. Brown, Bristol, Connecticut, c. 1845. Mahogany case with ripple base and beveled door, black and gold decorated upper glass and reverse painted lower glass with a view of Buckingham Palace.
Height 15 in. $600
Lindy Larson

Cornice Top Shelf Clock, Brinsmaid & Bros., Burlington, Vermont, c. 1840. Probably manufactured by C. & N. Jerome, Bristol, Connecticut. Mahogany case with round sides and cornice top, round brass dial is framed by a reverse painted glass, 30-hour brass movement.
Height 22 in. $600
Lindy Larson

Interior view of the J. C. Brown Rippled Cottage Clock, showing the unusual, compact 8-day movement; this movement was also used with J. C. Brown's fusee clocks.
Lindy Larson

Candlestick Clock, Terryville Manufacturing Company, Terryville, Connecticut, c. 1851. Black painted wood base and pedestal, signed paper dial, glass dome, one-day "pendulette"-style timepiece.
Height (incl. dome) 8 in. $900
Lindy Larson

Very Rare Early Shelf Clock, C. & N. Jerome, Bristol, Connecticut, c. 1838. Mahogany case with cornice top, two reverse painted glasses, round painted dial with brass hands, plain papered backboard. The 30-hour weight driven movement is noteworthy in that it was the first mass produced brass clock movement in America.
Height 22 in. $3000
Lindy Larson

A close-up of the interior of the C. & N. Jerome Shelf Clock; Jerome's patented movement revolutionized the clock making industry. Note this earliest model had solid brass wheels.
Lindy Larson

Early stenciled clock, J. C. Brown, Forestville, Connecticut, c. 1845. This particular style of clock was designed to be used either as a shelf, or a wall clock. Wood reverse ogee case painted black with stenciled and mother-of-pearl decoration, signed zinc dial is framed by a reverse painted upper glass, reverse painted lower tablet contains a portrait of J. C. Brown. 8-day spring driven brass movement.
Height 17 in. $1750
Lindy Larson

Interior view of the J. C. Brown Shelf/Wall Clock.
Lindy Larson

Rare Shelf Clock, F. & E. Sanfords, Goshen, Connecticut, c. 1818. Mahogany case with cornice top and satinwood inlay, reeded pillars, painted wood dial with false painted winding holes. One-day Torrington-style wood movement is not key wound, but of pull-up design like wood tall case movements.
Height 28 1/2 in. $5500
Lindy Larson

Interior view of the Sanfords Shelf clock, showing the unusual 4 arbor pull-up movement.
Lindy Larson

Interior view of the Aaron Crane Shelf Clock.
Lindy Larson

Shelf Clock, Aaron Crane, Newark, New Jersey, c. 1851. Figured mahogany case with a deeply etched glass tablet, glass dial reads, "J. R. Mills and Co., New York, A. D. Crane's Patent. " Originally designed to house a weight driven torsion pendulum, in the later years of production these cases were fitted with the more customary 8-day spring driven movement. Height 18 1/2 in. $3250
Lindy Larson

Close-up view of the unusual glass dial of the Aaron Crane Shelf clock.
Lindy Larson

Miniature Shelf Clock, Atkins, Whiting & Co., Bristol, Connecticut, c. 1850-1854. Mahogany case with beveled sides, stenciled decorations, reverse painted upper glass frames the dial and a reverse painted lower tablet of a locomotive, painted zinc dial, 30-hour time-and-strike movement.
Height 12 in. $1000
Lindy Larson

Interior view of the Miniature Shelf Clock by Atkins, Whiting & Co., showing the unusual movement with lyre plates and the Ives patented roller verge.
Lindy Larson

Four nice examples from left to right; on the left is a Mahogany Veneer Looking Glass Shelf Clock, Forestville Manufacturing Company, Forestville, Connecticut, c. 1845. It has a painted wooden dial, housing a brass scroll 8-day time-and-strike weight driven movement.
Height 30 in. $300

The second clock is an Ogee Double Door Shelf Clock, Waterbury, Connecticut, c. 1840. There is a floral decorated painted zinc dial, gilt transfer lower tablet, housing a brass weight driven 8-day time-and-strike movement. Minor veneer damage may be seen.
Height 29 in. $300

Clock number three is a Mahogany Veneer Ogee Shelf Clock, E. C. Brewster & Company, Bristol, Connecticut, c. 1861. The dial is floral decorated painted zinc, gilt transfer tablet incorporating two elk, housing a 30-hour spring driven time-and-strike movement, with a cast iron backplate.
Height 26 in. $400

The fourth clock is a Grain-painted Double Dial Calendar Clock, New Haven Clock Company for the National Calendar Clock Company, c. 1860. The top is corniced, black and gold dials, painted columns with gilt capitals and bases, ogee base, housing a brass spring driven 8-day time-and-strike movement, with simple calendar mechanism.
Height 27 in. $500
Robert W. Skinner, Inc.

From left to right:
Rare Shelf Clock, S. B. Terry, Terryville, Connecticut, c. 1840. "Round Side" with mahogany case, exceptional painted tablet, paper on wood dial inscribed "Silas B. Terry, Terry's Ville Connt," very unusual brass 30-hour time-and-strike movement retaining original weights and pendulum.
Height 22 1/2 in. $900

Mahogany Veneered Ogee Clock, S. B. Terry, Terryville, Connecticut, c. 1855. Mirror tablet, painted wood dial, paper label, 30-hour time-and-strike brass movement with iron weights and pendulum. Movement is a standard Connecticut movement, probably purchased by Terry and sold under his name.
Height 24 1/4 in. $200

Very Rare Shelf Clock, Silas B. Terry, Terryville, Connecticut, c. 1840. Mahogany veneered case with a molded door and very fine painted tablet with heart-shaped pendulum opening, original paper dial backed with wood inscribed "Silas B. Terry, Terry's Ville, Conn't," unusual and rare 30-hour weight driven brass time-and-strike movement retaining original iron weights and pendulum.
Height 24 in. $2750
Richard A. Bourne Co., Inc.

Two rare clocks, from left to right:
Rare Ogee Shelf Clock, unsigned - attributed to Joseph Ives, Plainville, Connecticut, c. 1842. Veneered with fancy mahogany, this unusual ogee features gilded columns inside the case and a painted zinc dial that creates a cover to the brass canister containing the movement. A very unusual 8-day time-and-strike weight driven movement has roller pinions, and has the count wheel mounted on the front of the front plate.
Height 30 in. $600
Rare Double Dial Calendar Ogee Clock, manufactured by the New Haven Clock Co. for the National Calendar Clock Co., New York, c. 1875. Rosewood veneered case, black paper dials with gold numerals, both inscribed "Manufactured for the National Clock Co. " 8-day movement striking on a "Cathedral" gong, simple calendar movement is driven by a lever off of the upper movement, and the pendulum is visible between the two dials.
Height 26 in. $950
Richard A. Bourne Co., Inc.

Three shelf clocks, from left to right:
Shelf Clock, Miles Lorse, Plymouth Hollow, Connecticut, c. 1855. Black painted wood case with mother-of-pearl inlay, excellent gold decoration, two painted tablets, the lower of Buckingham Palace, 8-day time-and-strike movement marked "Morse #1," painted zinc dial and paper instruction label.
Height 16 in. $400
Rare Shelf Clock, J. C. Brown, Bristol, Connecticut, c. 1855. Beautifully veneered rosewood case, two fine painted tablets, painted zinc dial with traces of the maker's name, paper label, and an 8-day time-and-strike movement marked "Forestville Manufacturing Co. " (a later name for the J. C. Brown Clock Co.). Signed on label, dial and movement.
Height 15 1/4 in. $400
Scarce Shelf Clock, New Haven Clock Co., New Haven, Connecticut, c. 1860. An unusual and early rosewood veneered case with gilded quarter columns, a painted glass, a painted zinc dial, maker's label, and an 8-day time-and-strike movement inscribed with the maker's name.
Height 14 1/2 in. $425
Richard A. Bourne Co., Inc.

Three shelf clocks, from left to right:
Shelf Clock, J. C. Brown Clock Co., Bristol, Connecticut, c. 1875. This clock in extraordinary condition features a highly figured rosewood case inscribed by the maker on the bottom molding, two nearly perfect painted tablets, painted zinc dial inscribed "J. C. Brown, Bristol, Ct., U. S.," perfect paper label, 8-day time-and-strike movement also inscribed by the maker.
Height 15 in. $1250
Shelf Clock, Chauncey Jerome, New Haven, Connecticut, c. 1860. Fine rosewood case with gilded columns, painted zinc dial, fine painted tablet of an eagle, paper label, 8-day time-and-strike movement.
Height 18 in. $325
Four Column Shelf Clock, New Haven Clock Co., New Haven, Connecticut, c. 1865. Mahogany case with four turned columns, paper dial, painted glass tablet and 8-day time-and-strike movement.
Height 19 in. $450
Richard A. Bourne Co., Inc.

Three assorted shelf clocks, from left to right:
Rare Fusee Shelf Clock, Chauncey Jerome, New Haven, Connecticut, c. 1860. In untouched original condition, this rosewood veneered shelf clock with gilded columns retains a perfect paper label, nearly perfect painted zinc dial, fine painted glass with gold decoration and a blue background. 8-day time-and-strike fusee movement inscribed by the maker.
Height 18 1/4 in. " $950
Rare Fusee Shelf Clock, Forestville Manufacturing Co., Bristol, Connecticut, c. 1850. Rosewood veneered case with applied ripple moldings, fine paper label, original etched glass tablet, painted zinc dial, 8-day time-and-strike fusee movement inscribed by the maker.
Height 18 in. $600
Rippled Shelf Clock, Forestville Manufacturing Co., Bristol, Connecticut, c. 1850. Rosewood veneered case with applied ripple moldings, frosted tablet, painted dial, paper label, 8-day time-and-strike movement inscribed "J. C. Brown. "
Height 17 in. $450
Richard A. Bourne Co., Inc.

Three shelf clocks, from left to right:
Very Rare Reverse Miniature Flat Ogee Shelf Clock, Henry Sperry & Co., New York, c. 1855. Grain decorated pine case, painted zinc dial with maker's name, exceptional painted tablet, 30-hour time movement.
Height 16 1/2 in. $725
Shelf Clock, "London" Model, Atkins Clock Co., Bristol, Connecticut, c. 1865. Rosewood veneered case with exceptional graining, two black and gold painted tablets, maker's label, 8-day time, strike-and-alarm movement.

Height 16 3/4 in. $625
"Venetian" Style Shelf Clock, E. Ingraham Clock Co., Bristol, Connecticut, c. 1875. Rosewood veneer and walnut round top case with a figure "8" door; in lower section of door is an original green and gold tablet, paper label and 8-day time-and-strike movement.
Height 18 in. $400
Richard A. Bourne Co., Inc.

Three shelf clocks, from left to right:
Ogee Clock, Terry & Andrews, Bristol, Connecticut, c. 1845. Mahogany veneered case with exceptional wood dial, decorated tablet, paper label, 30-hour time-and-strike brass movement with iron weights and period pendulum.
Height 26 in. $300
Scarce Sleigh Front Shelf Clock, Birge & Fuller, Bristol, Connecticut, c. 1845. Figured mahogany and mahogany veneered case with two excellent painted tablets, painted wood dial, "Puffin Betsy" paper label, and an 8-

day time-and-strike strap brass movement with roller pinions and period iron weights, as well as brass pendulum.
Height 33 in. $750
Split Column Shelf Clock, Boardman & Wells, Bristol, Connecticut, c. 1835. Mahogany and mahogany veneers with stenciled columns and splat, painted wood dial, tablet with a paper print, paper label, 30-hour time-and-strike wood movement with brass bushings.
Height 31 1/2 in. $275
Richard A. Bourne Co., Inc.

Two fine shelf clocks, from left to right:

Shelf Clock, Seth Thomas Clock Co., Thomaston, Connecticut, c. 1880. The "Garfield" Model in a solid oak case, retaining the original finish, painted zinc dial with the Seth Thomas trademark, paper maker's label, 8-day time-and-strike movement that strikes on a cathedral gong retaining original brass cased weights.
Height 29 1/2 in. $950

Rare Empire Shelf Clock, S. B. Terry, Plymouth, Connecticut, c. 1835. Magnificent mahogany case with turned columns, beautifully carved crest, two painted tablets of fine quality, maker's label, Terry's round time-and-strike movement, also of excellent quality and retaining two period weights.
Height 38 in. $950

Richard A. Bourne Co., Inc.

Three different shelf clocks, from left to right:

Empire Shelf Clock, S. B. Terry, Plymouth, Connecticut, c. 1855. Case of choice mahogany and mahogany veneers, painted tablet, painted wood dial, paper label, fine round time-and-strike weight driven brass movement.
Height 32 1/2 in. $750

Rare Shelf Clock, Silas B. Terry, Terry's Ville, Connecticut, c. 1840. Mahogany veneered case with a mirror tablet opening to a mint paper on wood dial inscribed by the maker, unusual 30-hour time-and-strike brass movement with a large escape wheel visible through the dial.
Height 24 1/4 in. $600

Rare Double Door Ogee Clock, movement: S. B. Terry; label: J. J. Beals, Boston, Massachusetts, c. 1840. Unusually large rosewood and mahogany veneered case featuring an exceptional wood dial with a brass seconds disc, paper label, high quality and unusual 8-day time-and-strike S. B. Terry movement retaining the original "gridiron" pendulum.
Height 34 1/2 in. $1150

Richard A. Bourne Co., Inc.

Three shelf clocks, from left to right:

Empire Shelf Clock, Henry Terry, Plymouth, Connecticut, c. 1830. In a mahogany case with three mirror tablets, gilded columns and cornice, retaining the original maker's label, excellent painted wood dial, 30-hour time-and-strike wood movement.
Height 31 1/2 in. $450

Transition Clock, Eli Terry & Sons, Plymouth, Connecticut, c. 1830. Mahogany case with stenciled splat and quarter columns, door with remains of original painted tablet, painted wood dial, paper label and 30-hour time-and-strike wood movement.
Height 28 1/2 in. $425

Shelf Clock, Silas Hoadley, Plymouth, Connecticut, c. 1835. Mahogany veneered case with stenciled quarter columns, fine glass in the crest with the inscription "Union Must Be Preserved," mirror tablet, painted wood dial, paper label, ivory bushed 30-hour wood movement retaining original iron weights.
Height 29 3/4 in. $750

Richard A. Bourne Co., Inc.

Three variations of shelf clocks, from left to right:

Very rare Miniature Ogee Clock, Chauncey Boardman, Bristol, Connecticut, c. 1845. A fine mahogany case with a mirror tablet, painted wood dial, rare 30-hour time-and-strike fusee movement with maker's mark.
Height 18 in. $700

Shelf Clock, Seth Thomas Clock Co., Thomaston, Connecticut, c. 1870. Known as the "Clubfoot" model with choice walnut veneers and moldings, painted zinc dial, maker's label and an 8-day time, strike-and-alarm movement.
Height 15 1/2 in. $375

Miniature Ogee Clock, Seth Thomas Clock Co., Thomaston, Connecticut, c. 1870. Walnut veneered case, stenciled tablet, 30-hour time-and-strike movement, paper label and painted zinc dial.
Height 16 1/2 in. $350

Richard A. Bourne Co., Inc.

Two rare shelf clocks, from left to right:
Atkins & Porter, Bristol, Connecticut, c. 1850. Excellent mahogany and rosewood veneered case, superb original painted tablet and zinc dial, 30-hour time-and-strike weight-driven movement.
Height 24 1/2 in. $425
Titus Roberts, Bristol, Connecticut, c. 1840. Made for Henry Hart, with a mahogany veneered case, paint decorated and gilded columns, painted wood dial, black and gold tablet, paper label, and a 30-hour time-and-strike weight-driven wood movement.
Height 32 3/4 in. $575
Richard A. Bourne Co., Inc.

From left to right are three clocks, the first being a Mahogany Veneer Transitional Shelf Clock, Norris North, Torrington, Connecticut, c. 1820. It has a painted and decorated wooden dial, reverse painted door tablet, baluster turned half columns, turned feet, housing a typical Torrington type 30-hour weight driven movement. Crest missing and some paint loss.
Height 26 in. $550
The middle clock is a Mahogany Veneer Ogee Shelf Clock, Ray & Ingraham, Bristol, Connecticut, c. 1840. It has a painted and decorated wooden dial, mirror tablet, 30-hour weight driven brass time, strike-and-alarm movement.

Height 25 3/4 in. $175
The clock on the right is a Mahogany Veneer Transitional Mantel Clock, Hopkins & Alfred, Harwinton, Connecticut, c. 1825. Incorporated are a stenciled crest, ebonized half columns, decorated and painted wooden dial, reverse painted door tablet (cracked), the whole raised on hairy paw feet, housing a wooden 30-hour time-and-strike movement. Dial and tablet paint flaking.
Height 28 1/4 in. $600
Robert W. Skinner, Inc.

Two choice shelf clocks, from left to right:
Scarce Shelf Clock, Ephraim Camp, Salem Bridge, Connecticut, c. 1830. Mahogany case with carved crest, capitals, and paw feet; painted wood dial with exceptional decorations, paper label, painted glass tablet, 8-day time-and-strike Salem Bridge movement, retaining iron weights and pendulum.
Height 31 in. $1750
Rare Shelf Clock, Silas B. Terry, Plymouth or Bristol, Connecticut, c. 1830. Early model with a quality mahogany veneered case, painted tablet, exceptional painted dial, and an unusual 8-day time-and-strike movement with heavy open work plates, maintaining power, rack-and-snail strike, seconds indicator, period iron weights and pendulum.
Height 27 1/2 in. $1950
Richard A. Bourne Co., Inc.

Three excellent shelf clocks, from left to right:
Half Pillar Shelf Clock, Seth Thomas Clock Co., Plymouth Hollow, Connecticut, c. 1860. Mahogany veneered case with paint decorated and gilded columns, exceptional painted tablet depicting "High Rock & Iodine Springs," painted zinc dial, paper label, 30-hour time-and-strike movement with period weights and pendulum.
Height 25 in. $295
Very Rare Shelf Clock, Seth Thomas Clock Co., Plymouth Hollow, Connecticut, c. 1865. A very scarce case model with patriotic motifs including the door tablet of a balloon and American flags, and a tablet in the arch of a spread eagle. Mahogany veneered case, painted zinc dial, paper label, 30-hour time-and-strike movement with iron weights and pendulum.
Height 29 in. $700
Shelf Clock, Chauncey Jerome, New Haven, Connecticut, c. 1850. Mahogany veneered case with grain-painted columns, two exceptional glass tablets, fine paper label, painted wood dial, 30-hour time-and-strike movement with period iron weights and pendulum.
Height 27 in. $500
Richard A. Bourne Co., Inc.

Three appealing shelf clocks, from left to right:
A Rare Sleigh-Front Shelf Clock, Silas B. Terry, Terryville, Connecticut, c. 1860. Fine rosewood veneered case with two appealing tablets, maker's label, unusual 8-day movement with count wheel strike and solid great wheels retaining iron weights and pendulum, painted zinc dial.
Height 32 1/2 in. $1400
Rare Shelf Clock, R. & J. B. Terry, Bristol, Connecticut, c. 1835. Produced by a company that was only in business for one year between 1835 and 1836, with a mahogany case, carved paw feet and eagle crest, painted decorated columns, paper label, painted wood dial, 8-day time-and-strike strap brass movement with iron weights and pendulum. Center tablet with stenciled eagle and lower door with fine painted tablet.
Height 38 in. $700
False Hollow Column Shelf Clock, Rodney Brace, North Bridgewater, Massachusetts, c. 1830. Mahogany and mahogany veneer Shelf clock with painted tablet, paper label, painted wood dial, 30-hour time, strike-and-alarm wood movement with iron weights and pendulum.
Height 33 in. $500
Richard A. Bourne Co., Inc.

From left to right:
Solid Walnut Shelf Clock, Seth Thomas, Thomaston, Connecticut, c. 1890. Painted zinc dial, 8-day time-and-strike movement, nickel plated pendulum with damascene decoration, two brass weights.
Height 29 1/2 in. $1400
In the middle, a Fashion Model #6 Calendar Clock, Seth Thomas, Thomaston, Connecticut, c. 1885. Solid walnut case, black and gold painted zinc dials, 8-day time-and-strike movement with long pendulum, three turned finials and door tablet with inscription "Fashion. "

Height 33 1/2 in. $1900
On the right, "Longbranch" Model Shelf Calendar Clock, William L. Gilbert Clock Co., Winsted, Connecticut, c. 1880. Solid walnut case with carved and turned decorations, door glass with silver decorations, paper-on-zinc dial, 8-day time-and-strike movement with simple calendar, and nickel and brass pendulum.
Height 27 in. $1075
Richard A. Bourne Co., Inc.

Shelf Clock, Seth Thomas Clock Company, Plymouth Hollow, Connecticut, c. 1868. Mahogany veneered case with half columns, painted tablet, enameled zinc dial, 8-day time-and-strike movement, with pendulum and iron weights. $500

Two examples of shelf clocks are, on the left, a Very Rare Empire Shelf Clock, Seth Thomas, Plymouth, Connecticut, c. 1835. This clock features a large mahogany and mahogany veneered case with exceptional carved and gilded Basket of Fruit on the crest, beautifully stenciled columns, and exceptional carved paw feet. The glazed door with original mirror opening to a perfect original black and gold dial and paper label. 30-hour time-and-strike wood movement, retaining original weights. Case sides veneered with highly figured mahogany.
Height 37 in. $1200
On the right is a Sleigh Front Shelf Clock, Seth Thomas, Plymouth Hollow, Connecticut, c. 1865. A mahogany veneered case with gilded columns, two painted tablets, enameled zinc dial, exceptional paper label, 8-day time-and-strike movement with pendulum and period iron weights.
Height 32 1/2 in. $500
Richard A. Bourne Co., Inc.

Three assorted shelf clocks, from left to right:
Tudor #1 Shelf Clock, Seth Thomas Clock Co., Thomaston, Connecticut, c. 1870. Round top case with highly figured rosewood and walnut in the original finish, painted zinc dial, fine paper label, 8-day time-and-strike movement inscribed by the maker with "S. T. " hands.
Height 15 in. $395
Flat top Shelf Clock, Seth Thomas Clock Co., Thomaston, Connecticut, Pat. 1863. Rosewood case with fine black and gold transferred tablet, painted zinc dial, maker's label, 30-hour time-and-strike lyre movement signed by the maker.
Height 15 1/2 in. $300
Shelf Clock - One of the City Series "New Orleans V. P. " Model, Seth Thomas Clock Co., Thomaston, Connecticut, c. 1875. Walnut case with clear glass, painted zinc dial, maker's label, 8-day time, strike-and-alarm movement inscribed by the maker.
Height 16 1/2 in. $300
Richard A. Bourne Co., Inc.

Three types of shelf clocks, from left to right:
Victorian Shelf clock, Seth Thomas Clock Co., Thomaston, Connecticut, c. 1870. Walnut case with carved decorations, glass with silver decoration, paper dial and an 8-day time-and-strike lyre movement.
Height 18 3/4 in. $175
Shelf Clock, Seth Thomas Clock Co., Thomaston, Connecticut, c. 1875. Rosewood veneered case with two gilded columns, pendulum viewing "bull's eye," maker's label, and an 8-day time-and-strike movement signed by the maker.

Height 16 in. $200
Scarce Victorian Shelf clock, Seth Thomas Clock Co., Thomaston, Connecticut, c. 1870. Walnut case with carved decorations, glass with gold designs, paper dial with the Seth Thomas trademark, alarm shut-off on top of clock, unusual time-and-alarm movement with the alarm mechanism in place of the strike train.
Height 21 1/2 in. $225
Richard A. Bourne Co., Inc.

Three different style shelf clocks, from left to right:
Rare "Venetian" Style Shelf Clock, attributed to N. L. Brewster, Bristol, Connecticut, c. 1860. Rosewood veneered case with turned columns and an excellent eglomise tablet of a gilded spread eagle, Joseph Ives 8-day time-and-strike "Tin Plate" movement with "Squirrel Cage" roller escapement and rolling pinions.
Height 17 1/2 in. $1950
Rare Shelf Clock, S. Hoadley, Plymouth, Connecticut, c. 1835. Mahogany veneered case with reverse bevel, eglomise panel, superb painted wood dial with floral decorations, paper label and Hoadley's unusual upside-

down 30-hour time-and-alarm wood movement, retaining iron weights, pendulum and key.
Height 21 in. $1100
Shelf Clock, Atkins Clock Co., Bristol, Connecticut, c. 1860. Exceptional case with rosewood figured veneers, turned split columns, three paint-decorated panels, paper label, painted dial and an 8-day time-and-strike movement.
Height 16 3/4 in. $650
Richard A. Bourne Co., Inc.

A group of three shelf clocks, from left to right:
Rare Steeple Clock, attributed to Roswell Kimberly, Ansonia, Connecticut, c. 1850. Very unusual rosewood veneered case design by an obscure maker. Painted zinc dial, paper label, 8-day time- and-strike movement. Height 20 1/2 in. $325
Split Column Shelf clock, Seth Thomas Clock Co., Plymouth Hollow, Connecticut, c. 1865. Maple veneered case with split columns, retaining original painted tablet, painted zinc dial, paper label, 30-hour time-and-strike brass movement.
Height 25 in. $300
Split Column Shelf clock, William L. Gilbert, Winsted, Connecticut, c. 1860. Rosewood veneered case with paint decorated columns, floral tablet, painted zinc dial, paper label and 30-hour time-and-strike brass weight driven movement.
Height 25 in. $300
Richard A. Bourne Co., Inc.

Three quality shelf clocks, from left to right:
Very Rare Calendar Shelf Clock, Gilbert Clock Co., Winsted, Connecticut, c. 1875. Standard "Sharp Top" rosewood veneered case with the very scarce feature of a calendar in the center of the moon phases. The three point indicator hand has a crescent pointer that indicates the day of the month, and two perpendicular pointers that point to the month. The movable disc in the center of the dial is manually adjusted. 8-day time-and-strike movement inscribed with the maker's name, paper on zinc dial, glass decorated with silver designs.
Height 17 1/2 in. $950
Rare Shelf Clock, labeled "From A. D. Smith's Clock Warehouse, E. 5th St.,
Cincinnati, Ohio. " Very unusual style rosewood veneered case with gilded split columns, paper on zinc dial, painted floral glass, an 8-day time-and-strike movement.
Height 17 in. $900
Shelf Clock, Seth Thomas Clock Co., Thomaston, Connecticut, c. 1875. Walnut veneered case with painted gold trim, a colorful glass with a bird in the center, painted zinc dial, paper label and an 8-day movement inscribed by the maker.
Height 15 1/2 in. $250
Richard A. Bourne Co., Inc.

Three assorted shelf clocks, from left to right:
Steeple Clock, O. H. Barton, Fort Edwards, New York, c. 1870. Rosewood veneered case, painted tablet, painted zinc dial, maker's label, 8-day time-and-strike movement.
Height 20 in. $400
Miniature Ogee Clock, Waterbury Clock Co., Waterbury, Connecticut, c. 1870. Mahogany and rosewood veneered case, painted zinc dial, floral tablet, paper label and 30-hour time-and-strike movement.

Height 19 in. $250
Shelf Clock, Waterbury Clock Co., Waterbury, Connecticut, c. 1875. Rosewood veneered case with a "figure 8" door, painted zinc dial, 30-hour time, strike-and-alarm movement.
Height 17 1/2 in. $200
Richard A. Bourne Co., Inc.

Three different shelf clocks, from left to right:
Peak Top Cottage Clock, Terry Clock Co., Thomaston, Connecticut, c. 1870. Rosewood case with excellent label and glass, 30-hour time-and-strike brass movement.
Height 18 in. $500
The Italian #2 Clock, E. N. Welch Manufacturing Co., Bristol, Connecticut, c. 1870. Rosewood case, 30-hour time-and-strike movement, brass club

escapement.
Height 14 in. $400
Steeple Clock, Brewster & Ingraham, Bristol, Connecticut, c. 1850. Mahogany case, brass springs, 30-hour brass time-and-strike movement.
Height 20 in. $500
Richard A. Bourne Co., Inc.

Three clocks, from left to right:
Rare Oversize Fusee Steeple Clock, Chauncey Jerome, New Haven, Connecticut, c. 1845. Excellent example in nearly mint condition of a scarce clock. Constructed with figured mahogany veneers, painted tablet depicting "St. Paul's Cathedral," painted zinc dial inscribed by the maker, 8-day time-and-strike double fusee movement with pendulum.
Height 23 in. $475
Rare Acorn Clock, J. C. Brown/Forestville Manufacturing Co., Forestville, Connecticut, c. 1855. Laminated case veneered with rosewood, painted tablet with Floral decorations and Geometric designs, paper label, painted zinc dial inscribed by the maker, 8-day time-and-strike double fusee movement with pendulum.
Height 19 in. $600
Steeple-on-Steeple Clock, John Birge, Bristol, Connecticut, c. 1850. Mahogany veneered case with two exceptional glasses, paper label, painted zinc dial, 8-day time-and-strike fusee movement with pendulum.
Height 26 1/2 in. $700
Richard A. Bourne Co., Inc.

Three clocks, from left to right:
Beehive Clock, Brewster & Ingraham, Bristol, Connecticut, c. 1850. Mahogany veneered case with cut glass tablet, painted zinc dial inscribed by the maker, paper label, 8-day time-and-strike movement with original brass springs and pendulum.
Height 19 in. $550
Scarce Steeple-on-Steeple Clock, William S. Johnson, New York, c. 1850 or possibly earlier. Mahogany veneered case with turned finials, 2 doors with fine original painted tablets opening to a label, and painted zinc dial. 8-day brass spring driven time-and-strike movement with pendulum.
Height 23 1/2 in. $600
Scarce Shelf Clock, E. Ingraham Clock Co., Bristol, Connecticut, c. 1865. "Huron" Model with rosewood veneered case, paper-on-zinc dial, paper label, 8-day time, strike-and-alarm movement with pendulum.
Height 16 in. $450
Richard A. Bourne Co., Inc.

Four different shelf clocks, from left to right.
Rare Shelf Clock, Atkins, Whiting & Co., with Joseph Ives Lever Spring Movement, c. 1855. Beautifully constructed rosewood veneered case with two doors, the upper with a fine black and gold tablet, the lower with a mirror; rare 30-day Ives Lever Spring time movement.
Height 17 1/2 in. $1450
Scarce Steeple Clock, Chauncey Boardman, Bristol, Connecticut, c. 1845. Figured mahogany veneered case with frosted tablet, painted zinc dial, paper label, rare 30-hour time-and-strike fusee movement inscribed "C. Boardman. "
Height 20 in. $425

Steeple Clock, Pomeroy & Parker, Bristol, Connecticut, c. 1855. Mahogany veneered case with fine original tablet, painted zinc dial, paper label and 8-day time-and-strike movement.
Height 19 3/4 in. $400
"Column" Model Shelf clock, Seth Thomas Clock Co., Thomaston, Connecticut, c. 1870. Rosewood veneered case with fine rosewood moldings, gilded columns, an opening for viewing the swing of the pendulum, painted zinc dial, paper label, 8-day time, strike-and-alarm "lyre" movement.
Height 16 in. $350
Richard A. Bourne Co., Inc.

A selection of wall clocks, from left to right:
Cottage Clock, unsigned, Connecticut, c. 1860. Pine case with original glass tablet, painted zinc dial and 30-hour time-and-alarm movement.
Height 10 in. $250
Cottage Clock, Ingraham Clock Co., Bristol, Connecticut, c. 1880. Rosewood veneered case with a decorated tablet, paper on zinc dial, 8-day time-and-strike spring movement with pendulum.
Height 13 1/4 in. $150
Cottage Clock, unsigned, Connecticut, c. 1870. Walnut case with ebonized moldings, paper on zinc dial, 30-hour time movement.
Height 11 in. $200
Steeple Clock, E. N. Welch, Forestville, Connecticut, c. 1865. Fine rose-

wood veneered case, painted tablet, zinc dial, paper label and 30-hour time, strike-and-alarm movement with pendulum.
Height 19 1/2 in. $225
Miniature Steeple Clock, E. N. Welch, Forestville, Connecticut, c. 1865. Mahogany veneered case with colorful floral tablet, painted zinc dial, paper label and 30-hour time movement.
Height 14 1/2 in. $200
Steeple Clock, E. N. Welch, Forestville, Connecticut, c. 1860. Mahogany veneered case with an exceptional cut glass tablet, paper on zinc dial, paper label and 8-day time-and-strike movement with pendulum.
Height 19 1/2 in. $250
Richard A. Bourne Co., Inc.

Three different shelf clocks, from left to right:
Miniature "Venetian" Shelf Clock, E. Ingraham & Co., Bristol, Connecticut, c. 1875. Walnut case with an excellent painted tablet of an Eagle and an American Shield backed with lime green, maker's label, painted zinc dial, 30-hour time and alarm movement.
Height 13 in. $275

Rare Fusee Steeple Clock With Rippled Door, Forestville Manufacturing Co., Bristol, Connecticut, c. 1850. Blonde mahogany case with a frosted glass, painted zinc dial inscribed in the top part by the maker, paper label, 8-day time and strike fusee movement.
Height 20 in. $950

Round Top Shelf Clock, Seth Thomas Clock Co., Thomaston, Connecticut, c. 1875. Rosewood veneered case with gold leaf trim around the glasses, painted zinc dial, paper label, superb blue and gold glass with an Eagle in the middle and "E Pluribus Unum," 8-day time and strike lyre movement inscribed by the maker with "S.T." hands.

Two rare examples, from left to right:
Very Rare Oversize Steeple Clock, Silas B. Terry, Terryville, Connecticut, c. 1845. Mahogany veneered case with a painted tablet, enameled wood dial and paper label, the dial has a seconds indicator and an opening to make the 3-inch vertical balance wheel visible, 30-hour time-and-strike movement with heavy strap brass plates and two large wooden fusee cones. An extremely rare and unusual clock!
Height 24 1/2 in. $8000

Very Rare Oversize Beehive Clock, Smith's Clock Establishment, 37 Bowery St., New York, c. 1840. Mahogany veneered case with exceptional cut glass tablet, paper label, painted wooden dial, 8-day time-and-strike movement with rack-and-snail strike, cast iron backplate and pendulum. An excellent example of a very rare clock!
Height 24 in. $1350
Richard A. Bourne Co., Inc.

Three different styles of shelf clocks, from left to right:
Very Rare Rippled Round Gothic Shelf Clock, movement by Waterbury Clock Co., Waterbury, Connecticut, c. 1875. Exceptional rosewood veneered case with applied rippled moldings, etched glass tablet, painted zinc dial, paper label of Waterbury Clock Co., 8-day time- and-strike movement with pendulum. This was originally made with a 30-hour fusee movement that was converted to an 8-day movement around 1875, probably by the Waterbury Clock Co., due to an over-pasted Waterbury label, Waterbury-type dial and Waterbury movement, all of which have been in place for over 100 years or more.
Height 20 in. $1400
Scarce Rippled Fusee Steeple Clock, J. C. Brown, Forestville, Connecticut, c. 1850. Rosewood veneered case with etched glass tablet, painted zinc dial inscribed by the maker, paper label, 8-day time-and-strike fusee movement with pendulum.
Height 20 in. $900
Scarce Rippled Beehive Clock, J. C. Brown, Forestville, Connecticut, c. 1850. Fine rosewood veneered case with rippled moldings, cut glass tablet, painted zinc dial, paper label, 8-day time-and-strike movement with pendulum.
Height 19 in. $1000
Richard A. Bourne Co., Inc.

These four clocks are, from left to right:
A Rosewood Veneer Mantel Clock, E. Ingraham, Bristol, Connecticut, c. 1860. This is the "Grecian" model, housing a brass 8-day time-and-strike movement.
Height 14 3/4 in. $295
The second clock is a Rosewood Veneer Mantel Clock, Atkins Clock Company, Bristol, Connecticut, c. 1860. This is the "Venetian" model with painted metal dial, reverse painted tablet incorporating a gilt star, housing a brass 8-day time-and-strike movement.
Height 16 in. $200

The third clock is a Miniature Mahogany Veneer Ogee Shelf Clock, Ansonia, Connecticut, c. 1850. It has a floral decorated painted metal dial, polychrome transfer tablet incorporating a bird and branch, housing an 8-day time-and-strike spring driven movement.
Height 18 1/2 in. $325
The clock on the right is a Mahogany Veneer Beehive Clock, Terry & Andrews, Bristol, Connecticut, c. 1840. It has a painted metal dial, acid finish cut glass tablet, housing a brass lyre 8-day time-and-strike movement.
Height 19 in. $300
Robert W. Skinner, Inc.

Three varieties of shelf clocks, from left to right:
Calendar Shelf Clock, E. N. Welch Manufacturing Co., Bristol, Connecticut, c. 1870. Fine carved and rosewood veneered case, 8-day time-and-strike B. B. Lewis calendar movement.
Height 19 1/4 in. $375
"Empress" Model Shelf Clock, E. N. Welch Manufacturing Co., Bristol, Connecticut, c. 1870. Rosewood veneered case, bird on branch glass, 30-hour time-and-strike brass movement.
Height 16 in. $275
Beehive Clock, Waterbury Clock Co., Waterbury, Connecticut, c. 1870. Rosewood veneered case, 8-day time, strike-and-alarm brass movement.
Height 18 3/4 in. $150
Richard A. Bourne Co., Inc.

Three fine shelf clocks, from left to right:
Rippled Beehive, E. N. Welch Manufacturing Co., Forestville, Connecticut, c. 1855. Rosewood veneered case with applied rippled moldings, beautifully painted tablet, paper label, painted zinc dial, 8-day time-and-strike movement inscribed by the maker.
Height 18 1/4 in. $1250
Rippled Front Steeple Clock, Forestville Manufacturing Co., Bristol, Connecticut, c. 1850. Rosewood veneered case with applied rippled moldings, painted glass, painted zinc dial, maker's label on back board, 8-day time-

and-strike movement.
Height 19 3/4 in. $650
Rippled Beehive Clock, E. N. Welch Manufacturing Co., Forestville, Connecticut, c. 1855. Rosewood veneered case retaining an old darkened finish, a fine etched tablet with a heart-shaped design, painted zinc dial, maker's label and an 8-day time, strike-and-alarm movement.
Height 19 in. $650
Richard A. Bourne Co., Inc.

Three different shelf clocks, from left to right:
Very Rare Fusee Shelf Clock, New Haven Clock Co., New Haven, Connecticut, c. 1860. "Sharp Top" mahogany veneered case with a floral tablet, painted zinc dial, maker's label, and a very rare and unusual 8-day time-and-strike movement with fusee cones and spring barrels; movement is believed to have been designed by Hiram Camp, nephew of Chauncey Jerome. A very scarce example.
Height 17 in. $700
Very Rare Shelf Clock, Atkins Clock Co., Bristol, Connecticut, c. 1875. One of the most unusual and rare of the Atkins Clock Co. case styles, the clock has an oversized dial, a turned wood bezel and is veneered with choice

rosewood veneers, 8-day time-and-strike movement, fine paper label and painted zinc dial. Said to be one of two known.
Height 21 in. $1850
Shelf Clock, E. Ingraham & Co., Bristol, Connecticut, c. 1875. A variation of the mosaic front with light and dark alternating wood on the columns of the case, rosewood veneered case, paper on zinc dial, gold tablet with balloons and American flags, 30-hour time-and-strike movement.
Height 16 in. $650
Richard A. Bourne Co., Inc.

Six interesting clocks, from the top, left to right:
A Rare Miniature Steeple Clock, J. C. Brown, Forestville, Connecticut, c. 1865. Mahogany veneered case, exceptional glass tablet depicting a village scene, painted zinc dial inscribed by the maker, paper label, 30-hour time-and-strike movement.
Height 16 in. $500
Very Rare Miniature Rippled Beehive Clock, J. C. Brown, Forestville, Connecticut, c. 1865. Rosewood veneered case with applied rippled moldings, painted floral tablet, painted zinc dial inscribed by J. C. Brown, paper label, 30-hour time-and-strike movement with pendulum.
Height 15 1/2 in. $2000
Very Rare Miniature Rippled Steeple Clock, Forestville Manufacturing Co., Bristol, Connecticut, c. 1865. Rosewood veneered case with applied ripple moldings, patriotic tablet with cannons and flags, painted zinc dial, paper label, 30-hour time-and-strike movement with pendulum.
Height 16 in. $1300
Rosewood Veneered Steeple Clock, Brewster & Ingraham, Bristol, Connecticut, c. 1848. Exceptional cut glass tablet, painted zinc dial inscribed by the maker, paper label, 8-day time-and-strike movement with cast iron main spring barrels and repeating strike mechanism.
Height 20 in. $650
In the middle, a Rosewood Veneered Beehive Clock, Brewster Manufacturing Co., Bristol, Connecticut, c. 1852-1856. Cut glass tablet, painted zinc dial inscribed by the maker, paper label, 8-day time-and-strike movement with pendulum.
Height 19 in. $600
Lastly, Kirk's Patent Steeple Clock, Brewster & Ingraham, Bristol, Connecticut, c. 1850. Mahogany veneered case with painted tablet, painted zinc dial, maker's label, 30-hour time-and-strike movement with cast iron backplate patent by Charles Kirk. Retains pendulum, movement is inscribed by hand, "Brewster & Ingraham, Bristol, Ct."
Height 20 in. $375
Richard A. Bourne Co., Inc.

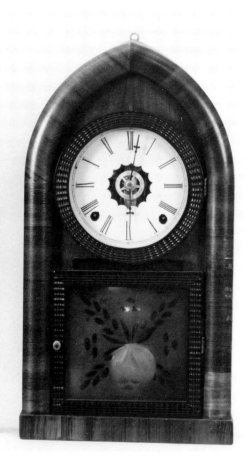

Three shelf clocks, from left to right:
Rare Ripple Beehive Clock, Forestville Manufacturing Co., Bristol, Connecticut, c. 1850. Rosewood veneered case with an exceptional etched glass tablet, painted zinc dial, maker's label, 8-day time-and-strike movement inscribed by the maker.
Height 19 in. $750
Rippled Front Steeple Clock, Forestville Manufacturing Co., Bristol, Connecticut, c. 1850. Mahogany veneered case with applied rippled moldings, painted tablet with flowers and a bird, painted zinc dial, 8-day time, strike-and-alarm movement inscribed by the maker.
Height 19 3/4 in. $675
Richard A. Bourne Co., Inc.

Three styles of shelf clocks, from left to right:
Rare Ripple Beehive Clock, J. C. Brown, Bristol, Connecticut, c. 1850. Rosewood veneer case with applied ripple molding, in the door is an exceptional cut glass tablet with a heart-shaped design; the dark rosewood rippled molding is of unusually fine quality. The rare feature is an 8-day time-and-strike movement with alarm.
Height 19 in. $1200
Steeple Clock With Rippled Door, Brewster & Ingraham, Bristol, Connecticut, c. 1850. Rosewood veneered case with exceptional cut glass tablet, painted zinc dial, paper label with over-painted label of Daniel Pratt Jr., Boston. 8-day time-and-strike movement retaining the original brass springs, and inscribed by the maker.
Height 20 in. $625
Ripple Beehive Clock, J. C. Brown, Bristol, Connecticut, c. 1850. Rosewood veneer case, unusual type of ripple molding, uncommon design on the etched tablet, paper label, 8-day time-and-strike movement inscribed by the maker.
Height 19 in. $1500
Richard A. Bourne Co., Inc.

Three assorted shelf clocks, from left to right:
Shelf Clock, Seth Thomas Clock Co., Thomaston, Connecticut, c. 1870. Known as the "Club Foot" Model with choice walnut veneers and moldings, painted zinc dial with a Seth Thomas trademark, maker's label, 8-day time-and-strike movement.
Height 15 1/2 in. $550
"Venetian" Shelf Clock, E. Ingraham Clock Co., Bristol, Connecticut, c. 1875. An unusual variation of the standard case style, with rosewood veneer, exceptional gold painted glass, paper on zinc dial, 8-day time-and-strike movement marked with the maker's name.
Height 17 1/2 in. $450
Shelf Clock, New Haven Clock Co., New Haven, Connecticut, c. 1870. Rosewood veneered case with applied gilded columns, fine tablet, painted zinc dial, 8-day time-and-strike movement inscribed by the maker.
Height 16 in. $500
Richard A. Bourne Co., Inc.

Three scarce shelf clocks, from left to right:
"Oriental" Style Shelf Clock, E. Ingraham & Co., Bristol, Connecticut, c. 1865. Rosewood veneered case, gilded capitals, paper-on-zinc dial, 8-day time, strike-and-alarm movement with pendulum.
Height 18 in. $650
In the middle, a scarce "Venetian Mosaic Front" Shelf Clock, E. Ingraham & Co., Bristol, Connecticut, c. 1870. Light and dark laminated wood on the columns, paper-on-zinc dial, gold-leaf tablet depicting a Bird, paper label, 8-day time-and-strike mechanism with pendulum.
Height 15 3/4 in. $450
On the right, a scarce "Dakota" Model Shelf Clock, E. Ingraham & Co., Bristol, Connecticut, c. 1870. Rosewood veneered case, zinc dial, paper label, 8-day time-and-strike movement with pendulum.
Height 15 3/4 in. $600
Richard A. Bourne Co., Inc.

Empire Shelf Clock, Chauncey Jerome, Bristol, Connecticut, c. 1860. Unusual flat front case in light colored bird's-eye maple veneer, with the door a darker figured contrasting mahogany veneer, painted wood dial, reverse painted tablet, 30-hour weight driven brass movement of typical ogee style. Height 24 in. $600
Lindy Larson

Interior view of the C. Jerome Empire Shelf Clock.
Lindy Larson

Shelf Clock, Elisha Manross, Bristol, Connecticut, c. 1845. Rosewood veneered case with lighter inlay, painted metal dial, tablet of "Ballston Springs," 8-day time-and-strike movement.
Height 14 in. $500
American Clock & Watch Museum

Shelf Clock, Seth Thomas Clock Company, Thomaston, Connecticut, c. 1890. Oak case, pedestal base. $400

Shelf Clock, Chauncey Boardman, Bristol, Connecticut, c. 1835. Mahogany veneered beveled case, paper on zinc dial, exquisitely painted tablet with patriotic motifs, marked "Boardman & Wells, Bristol, Ct. " on the bottom face of tablet, 30-hour wooden time-and-strike movement.
Height 26 3/4 in. $1000
American Clock & Watch Museum

Ornate Shelf Clock, E. Ingraham & Co., Bristol, Connecticut, c. 1885. Gold painted oak case, ornate decoration, painted zinc dial, 8-day time-and-strike brass movement.
Height 18 1/2 in. $600
American Clock & Watch Museum

Round Gothic Four Poster Shelf Clock, Terry & Andrews, Bristol, Connecticut, c. 1845. Mahogany case with four turned columns and finials, cut glass tablet, painted zinc dial.
Height 19 1/2 in. $1100
Lindy Larson

Interior view of the Terry & Andrews Gothic Four Poster Shelf clock, showing the signed 8-day lyre movement and separate alarm movement typically used by this firm. Often these movements were equipped with large brass mainsprings.
Lindy Larson

Round Gothic Four Poster Shelf Clock, Brewster & Ingraham, Bristol, Connecticut, c. 1845. Rosewood case with four finely turned finials and full columns, stenciled and reverse painted glass, painted metal dial. Height 19 1/2 in. $1000
Lindy Larson

Interior view of the Round Gothic Four Poster Shelf Clock by Brewster & Ingraham, showing the signed 8-day movement; the plates of the movement have a raised rib, presumably for added strength.
Lindy Larson

"Roman" Model Shelf Clock, E. N. Welch Manufacturing Co., Forestville, Connecticut, c. 1870. Rosewood case with rosewood graining on the door, painted zinc dial, 30-hour time-and-strike movement.
Height 15 in. $450
Lindy Larson

"Tuscan" Model Shelf Clock, E. Ingraham & Co., Bristol, Connecticut, c. 1875. Mahogany veneered case, painted tablets, paper on zinc dial, 8-day brass time-and-strike movement.
Height 18 1/2 in. $500
American Clock & Watch Museum

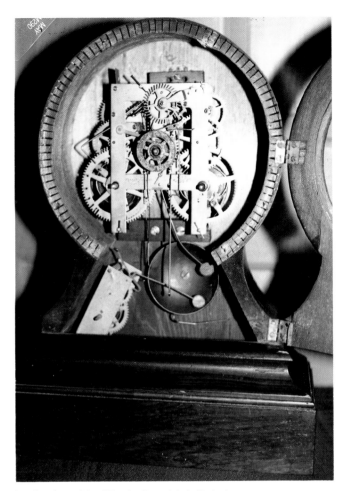

Interior view of the "Grecian" model shelf clock.
Lindy Larson

"Grecian" Model Shelf Clock, E. N. Welch Manufacturing Co., Bristol, Connecticut, c. 1870. Rosewood case with rosewood graining on the door, carved leaves between the bezel and the base, painted zinc dial, 8-day time-and-alarm movement.
Height 15 in. $500
Lindy Larson

Rear view of the "Grecian" model shelf clock, showing the label.
Lindy Larson

Shelf Clock, E. Ingraham & Co., Bristol, Connecticut, c. 1880. Classic rounded rosewood veneered case, painted dial, simulated mercury pendulum, 8-day time-and-strike brass movement.
Height 16 in. $350
American Clock & Watch Museum

"Tuscan" Model Shelf Clock, E. Ingraham & Co., Bristol, Connecticut, c. 1870 Rosewood case of very unusual design and a very rare model. The two smaller round glasses were reverse painted; the two pictured here are replacements. The case is rosewood veneered and the door is grain-painted to imitate rosewood;, 8-day time-and-strike spring movement.
Height 18 1/2 in. $1450
Lindy Larson

"Grecian Mosaic" Model Shelf Clock, E. Ingraham & Co., Bristol, Connecticut, c. 1885. Beautiful alternating oak and mahogany veneered case, paper dial, 8-day brass spring driven movement.
Height 15 1/2 in. $600
American Clock & Watch Museum

Interior view of the "Tuscan" model shelf clock, showing the unusual door arrangement.
Lindy Larson

"Oriental" Model Shelf Clock, E. Ingraham & Co., Bristol, Connecticut, c. 1880. Rosewood veneered case with ornate carvings, painted zinc dial, 8-day time-and-alarm movement.
Height 18 in. $500
American Clock & Watch Museum

Shelf Clock, C. & N. Jerome, Bristol, Connecticut, c. 1835. Mahogany veneered pine case, rounded sides, door with curved molding follows the curve of the side. Curved cornice on top, both tablets painted green with beautifully applied floral decorations, 30-hour time-and-strike movement.
Height 22 in. $1150
American Clock & Watch Museum

Shelf Clock, Crane's Patent, New York, 1841. Grain-painted mahogany veneered case, painted metal dial, etched glass tablet, 30-day swinging ball, torsion escapement.
Height 19 3/4 in. $600
American Clock & Watch Museum

"Chinese Chippendale" Style Shelf clock, W. F. Barker, Newport, Rhode Island, c. 1820. Mahogany case, painted zinc dial with seconds indicator, 4-day weight driven time-and-strike brass movement, bolt-and-shutter maintaining power is activated by lever inside which opens shutter over the keyhole and applies power to the clock while being wound.
Height 21 3/4 in. $1250
American Clock & Watch Museum

Interior view of Crane's Patent Shelf Clock.
American Clock & Watch Museum

Shelf Clock, Brewster & Ingraham, Bristol, Connecticut, c. 1845. Mahogany veneered case, painted zinc dial, painted tablet of the "Monument at Lexington," 8-day time, strike-and-alarm movement.
Height 23 1/2 in. $650
American Clock & Watch Museum

Prototype Shelf Clock, Brewster & Ingraham, Bristol, Connecticut, c. 1850. Large elaborate shelf clock with rosewood veneered case, nine finials, painted metal dial, etched lower tablet, 8-day time-and-strike movement.
Height (to top of finial) 31 in. $1200
American Clock & Watch Museum

Shelf Clock, Atkins, Whiting & Co., Bristol, Connecticut, c. 1850. Rosewood veneered case, painted zinc dial with lovely floral surround in the upper tablet, mirror tablet in the bottom, 30-day time-only wagon spring movement.
Height 18 in. $600
American Clock & Watch Museum

Violin Clock, Seth Thomas Clock Co., Thomaston, Connecticut, c. 1885. Walnut case in the shape of a violin, painted metal dial, black tablet with gold decorated musical motifs, 8-day brass time-and-strike movement. Height 32 in. $1300
American Clock & Watch Museum

"Hour Glass" Style Shelf Clock, Joseph Ives, Plainville, Connecticut, c. 1850. Graceful mahogany veneered case with full mahogany columns, brass dial, 30-hour brass movement, stamped plates, roller pinion, fusee single hoop or wagon spring. Height 23 in. $1200
American Clock & Watch Museum

"Meditation" Model Mantel Clock, Seth Thomas Clock Co., Thomaston, Connecticut, c. 1910. White metal case with bronze plating and a gilded center panel, relief decorations of a victorian lady and flowers, separate porcelain numerals on a 4 inch gilt dial; this model was also made with a full porcelain dial. 8-day movement with cathedral gong.
Height 11 1/2 in. $400
Lindy Larson

Extremely Rare Lyre Clock, unknown, Massachusetts, probably Boston, c. 1820. Extraordinarily large and unique mahogany lyre clock, with a full-bodied carved and gilded eagle, beautifully carved throat with figured mahogany panel, lower door frame of half-round mahogany with a figured panel, and a glazed wood bezel opening to a mint original dial. Extremely rare and unusual 8-day time-and-strike movement with skeletonized plates striking on a belt inside the hood, and retaining the original lead weights.
Height 43 in. $4500
Richard A. Bourne Co., Inc.

Papier-m>chJ Mantel Clock, Litchfield Manufacturing Co., Litchfield, Connecticut, c. 1855. Papier-m>chJ case stenciled, painted and inlaid with mother-of-pearl on a scroll base with brass bezel sash, painted metal dial with heavy floral decoration, unique 8-day brass marine movement with horizontal balance assembly and peculiar escapement probably by Matthews and Jewel of Bristol, Connecticut.
Height 21 in. $350
American Clock & Watch Museum

Papier-m>chJ Mantel Clock, Alfred Bennett, Portsmouth, New Hampshire, c. 1868. Papier-m>chJ case with mother-of-pearl inlay and painted decorations, two love birds kissing on the top of the clock, 30-hour brass marine type movement.
Height 18 in. $600
American Clock & Watch Museum

Papier-m>chJ Mantel Clock, C. Jerome, New Haven, Connecticut, c. 1855. Scrolled front case of papier-m>chJ, painted, gold leafed and inlaid with mother-of-pearl, scrolled molded base, brass bezel with 51/2 inch chapter on metal painted dial, 8-day brass time-and-strike movement.
Height 21 in. $450
American Clock & Watch Museum

Miniature Mantel Timepiece, Jerome Manufacturing Co., New Haven, Connecticut, c. 1853. Pressed brass case with porcelain dial, "Paris" is stamped into the case below the dial; wood stand with gilt stenciling and three brass feet, glass dome with rope trim at the bottom; this model was also made with a papier-mâché case. Unusual 2-day marine escapement designed by S. N. Bottsford.
Height (incl. dome) 11 1/2 in. $1000
Lindy Larson

Rear view of Miniature Mantel Timepiece; note the unusual shaped plates and the rear winding arbor.
Lindy Larson

"La Plata" Model Mantel Clock, Ansonia Clock Co., Ansonia, Connecticut, c. 1900. China case, heavily decorated, porcelain dial, exposed Brocot-style escapement, 8-day time-and-strike movement.
Height 13 in. $500
American Clock & Watch Museum

Heavily decorated China cased Mantel Clock, Ansonia Clock Company, Ansonia, Connecticut. Porcelain dial, 8-day movement. $450

"La Vera" Model Royal Bonn China Mantel Clock, Ansonia Clock Co., New York, c. 1900. Signed Royal Bonn case with Dutch scenes and rich brown trim, signed porcelain dial with open escapement and beveled glass, 8-day movement.
Height 13 1/2 in. $600
Lindy Larson

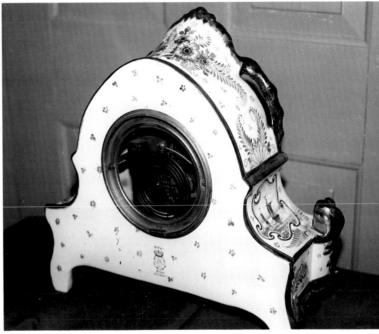

Rear view of the "La Vera" Mantel Clock, showing the Royal Bonn logo.
Lindy Larson

"Modena" Model Mantel Clock, Ansonia Clock Co., New York, 1905. Solid oak case, porcelain dial with brass center and beveled glass, ornate pressed designs and scrolled front, 8-day movement.
Height 17 1/2 in. $475
Lindy Larson

Three clocks, from left to right:
Double Dial Calendar Clock, Horton's patent by the Ithaca Clock Co., Ithaca, New York c. 1870. "Cottage" style case, original backboard with label, 8-day time, strike and calendar movement with club escapement, all in excellent condition.
Height 22 1/4 in. $325
Lincoln Model Shelf Clock, Seth Thomas Clock Co., Thomaston, Connecticut, c. 1880. Solid walnut Eastlake case, 8-day weight driven brass time-and-strike movement. Really excellent example.
Height 26 1/2 in. $395
Shelf Calendar Clock, Seth Thomas Clock Co., Thomaston, Connecticut, c. 1880. Lovely veneered walnut case, original instruction label, 8-day time, strike and calendar movement.
Height 26 3/4 in. $400
Richard A. Bourne Co., Inc.

Three attractive shelf clocks, from left to right:
Very Rare Steeple Clock, Roswell Kimberly, Ansonia, Connecticut, c. 1850. An unusual and unique case design with the label of Roswell Kimberly, an obscure and little known maker. Inscription on the top part of the case reads, "Patented Oct 11th 1850, by R. Kimberly, Ansonia, Ct. " Silver and black painted tablet of an eagle. 8-day time-and-strike movement.
Height 20 1/2 in. $950
Violin Clock, unsigned, c. 1890. Beautiful carved mahogany case in the shape of a violin, on a base, paper on zinc dial, black and gold glass. Name of maker is thought to be Fitzgerald. 8-day time-and-strike movement.
Height 28 in. $800
Rare Venetian #2 Mosaic Front Shelf Clock, E. Ingraham & Co., Bristol, Connecticut, c. 1875. Known as a "Candy Striper," this clock has alternating light and dark woods on the case front, a fine original tablet, paper on zinc dial, 8-day time-and-strike movement.
Height 18 in. $450
Richard A. Bourne Co., Inc.

Three rare shelf clocks, from left to right:
Rare Novelty Clock, Seth Thomas Clock Co., Plymouth Hollow. Connecticut, c. 1860. Alarm mechanism by "G. K. Proctor & Co., Beverly, Mass. " Rosewood veneered case, painted zinc dial, mirror tablet, 30-hour time-and-alarm movement. A very interesting and rare novelty clock with a mechanism on top that ignites a small lamp at the moment the alarm strikes.
Height 11 1/2 in. $1000
Rare Miniature Reverse Ogee Shelf Clock, Smith & Goodrich, Bristol, Connecticut, c. 1840. Unusual case veneered with rosewood, a fine painted and stenciled tablet, painted zinc dial, maker's label and a rare 30-hour time-and-strike fusee movement.
Height 15 in. $750
Rare Miniature Cottage Clock, Welch, Spring & Co., Forestville, Connecticut, c. 1865. Reeded walnut case, paper label, two black and gold glasses, paper on zinc dial, maker's signature on movement and label, 30-hour time-and-strike movement.
Height 11 in. $725
Richard A. Bourne Co., Inc.

Three beautiful shelf clocks, from left to right:
No. 3 1/2 Ithaca Parlor Calendar Clock, Ithaca Clock Co., Ithaca, New York, c. 1875. Beautiful walnut case with carved decoration, black and silver hour dial, glass silver and gold calendar dial, cut glass pendulum bob, maker's label, silvered hands, 8-day time-and-strike movement. Height 20 1/2 in. $1850
"Fashion Model #3" Calendar Clock, Southern Calendar Clock Co., St. Louis, Missouri, Pat. 1879. Movement by Seth Thomas Clock Co., Thomaston, Connecticut. Beautiful, solid walnut case with excellent moldings, columns and finials; door glass with original gold leaf inscription, paper label, 8-day time-and-strike movement with a separate calendar mechanism and dial. Height 32 in. $2050
Rare Solar Clock, Manufactured by Dr. Lewis Whiting, Saratoga, New York, patent by Theodore Timby, Baldwinsville, New York., c. 1863. Walnut case with an 8-day movement and a globe made and installed by the Boston firm of Gilman Joslin. Clock features a 12-hour dial above the globe and a minute dial below. The movement has a jeweled balance wheel; it is believed that only 600 of this style were ever made.
Height (to top of finial) 27 in. $3100
Richard A. Bourne Co., Inc.

Two unusual shelf clocks, from left to right:
Rare Solar Clock, Dr. Lewis Whiting, Saratoga, New York; patent by Theodore Timby, Baldwinsville, New York. c. 1880. Walnut case with a globe made and installed by the Boston firm of Gilman Joslin; features a 12-hour dial above the globe and a minute dial below, 8-day lever movement with a jeweled balance wheel. It is believed that only 1,000 of this type were produced.
Height 28 in. $2600
Victorian "Mirror Side" Mantel Clock, Ansonia Clock Co., Ansonia, Connecticut, c. 1880. An intricate carved oak case with two brass cherubs and two applied brass masks; behind the cherubs are mirror tablets. Retains the original silver decorated tablet, paper on zinc dial and an 8-day time, strike-and-alarm movement with pendulum and key.
Height 24 1/2 in. $750
Richard A. Bourne Co., Inc.

From left to right, three fine clocks:
Calendar Clock, Seth Thomas, Thomaston, Connecticut, c. 1877. Fashion #3, sold by Southern Calendar Clock Co., St. Louis, Mo. with walnut case, enameled zinc dials, the upper with seconds indicator and "Patented Dec. 28, 1875" and the lower with day and month apertures and "Southern Calendar Clock Co., patented March 18, 1879. " Glass tablet with gold inscription "Fashion," black and gold paper label, 8-day time-and-strike movement with pendulum, 3 turned finials.
Height 31 1/4 in. $550
Rare Shelf Clock, Seth Thomas, Thomaston, Connecticut, c. 1900. Scarce spring driven model with solid walnut case, enameled zinc dial inscribed "Seth Thomas," 8-day time-and-strike spring driven movement with pendulum. Paper label in bottom of case.
Height 29 1/2 in. $500
Calendar Clock, Seth Thomas, Thomaston, Connecticut, c. 1876. Fashion #8 with walnut case, enameled upper dial with seconds indicator and "Patent Dec. 28, 1875," paper on zinc calendar dial inscribed "Made by Seth Thomas Clock Co., Patented Feb. 15, 1876," with day and month apertures, 8-day time-and-strike movement with nickel plate damascene decorated pendulum bob.
Height 32 in. $750
Richard A. Bourne Co., Inc.

Four assorted shelf clocks, from left to right:
"No. 9 Shelf Cottage" Calendar Clock, Ithaca Calendar Clock Co., Ithaca, New York, c. 1870. Solid walnut case, paper on zinc dials, the lower inscribed by the maker and "Patented August 28, 1866," case fitted with an alarm mechanism in the bottom section, 8-day time-and-strike movement by E. N. Welch, calendar mechanism indicating day, month and date.
Height 22 1/4 in. $450
Scarce Shelf Clock, New Haven Clock Co., New Haven, Connecticut, c. 1875. Oak case with carved and pressed decorations, painted zinc dial, unusual 8-day fusee time-and-strike movement, believed to have been designed by Hiram Camp.
Height 23 1/2 in. $400
Scarce Steeple Clock, Chauncey Boardman, Bristol, Connecticut, label of William S. Johnson, New York dealer, c. 1845. Mahogany veneered case with green painted glass, painted zinc dial, paper label and a 30-hour time-and-strike fusee movement.
Height 19 1/2 in. $250
Shelf clock, F. Kroeber, New York, c. 1870. Rosewood veneered case with fine black and gold dial, brass hands, brass and beveled glass pendulum with fast/slow regulating gauge, 8-day time-and-strike movement.
Height 17 in. $200
Richard A. Bourne Co, m Inc.

An assortment of shelf and mantel clocks, from top to bottom, left to right:
Cottage Clock, Terhune & Edwards, New York, c. 1870. Walnut veneered case with a black and gold tablet, painted zinc dial, paper label, 30-hour time movement.
Height 12 in. $100
"Greek Model" Shelf Clock, Ansonia Clock Co., New York, c. 1880. Ebonized case with white decorations, fine paper label, paper on zinc dial and an 8-day time-and-strike movement.
Height 17 1/4 in. $150
Mantel Clock, Ansonia Clock Co., Ansonia, Connecticut, c. 1880. Covered with red velvet with applied white metal feet and decorations, porcelain dial, 8-day time-and-strike movement with visible escapement.
Height 13 in. $200
Shelf Clock, Sessions Clock Co., Forestville, Connecticut, c. 1890. Maple case with tiger maple front, paper dial, maker's label, 8-day time-and-strike movement.
Height 10 in. $175
Mantel Clock, Sessions Clock Co., Forestville, Connecticut, c. 1900. Ebonized case with marbleized columns and cast white metal feet, paper dial, 8-day time-and-strike movement.
Height 11 1/2 in. $150
Richard A. Bourne Co., Inc.

Six styles of shelf clocks, from left to right, top to bottom:
Scarce Cottage Clock, S. B. Terry Co., Plymouth, Connecticut, c. 1860. Mahogany and pine case, excellent painted wood dial, door retains original tablet, 30-hour "ladder" time-and-alarm movement.
Height 11 1/4 in. $650
Miniature Cottage Clock, Jerome & Co., New Haven, Connecticut, c. 1865. Superb rosewood veneered case, the "Rose" model with two painted tablets, paper label, painted zinc dial, 30-hour time-and-alarm movement.
Height 9 1/2 in. $300
Rare Miniature Cottage Clock, The Terry Clock Co., Waterbury, Connecticut, Patented 1868. Solid walnut case, painted zinc dial, 30-hour time-and-alarm movement inscribed by the maker, "The Terry Clock Co., Waterbury, Conn, Patented Dec. 1, 1868. "
Height 10 1/4 in. $350
Scarce Miniature Cottage Clock, Bradley & Hubbard, West Meriden, Connecticut, c. 1865. Rosewood veneered shelf clock with a mirror in the door, paper on zinc dial, maker's label, and a 30-hour time movement.
Height 9 in. $250
Scarce "Cigar Box" Clock, unsigned, Connecticut, c. 1860. Small pine clock with a mahogany veneered door, painted zinc dial, and a 30-hour time movement.
Height 9 1/2 in. $225
Scarce Box Clock, unsigned, Connecticut, c. 1860. Simple mahogany veneered case with an exceptional red and gold tablet, painted zinc dial, 30-hour time movement with the pendulum suspended from the top board. Possibly by the New England Clock Co., Bristol, Connecticut.
Height 11 3/4 in. $250
Richard A. Bourne Co., Inc.

Three different shelf clocks, from left to right:
Flat Top Shelf clock, Seth Thomas Clock Co., Plymouth, Connecticut, c. 1870. Rosewood veneered case, transferred tablet of two swans, painted zinc dial, maker's label, 8-day time-and-strike movement.
Height 13 3/4 in. $375
Rare Shelf Clock, Forestville Manufacturing Co., Bristol, Connecticut, c. 1860. Rosewood veneered case with gilded decorations, two painted tablets, maker's label and 8-day time-and-strike movement signed by the maker.
Height 15 in. $250
Shelf Clock, Seth Thomas Clock Co., Thomaston, Connecticut, c. 1875. Rosewood veneered case, painted black and gold tablet, painted zinc dial, maker's label, 30-hour time, strike-and-alarm movement.
Height 14 1/2 in. $250
Richard A. Bourne Co., Inc.

Three shelf clocks, from left to right:
Rare Cottage Clock, Chauncey Jerome, New Haven, Connecticut, C. 1850. Pine case, painted black with gilt decorations, painted wood dial, unusual 30-hour time movement inscribed by the maker with an upside-down verge, retaining the original brass main spring.
Height 12 in. $200
Rare Rippled Cottage Clock, William B. Lorton, New York, c. 1854. Walnut case with applied rippled moldings, paper label, paper on zinc dial, 30-hour time-and-strike movement, alarm mechanism winds through a hole in the front of the case. On the back of the clock is an 1867 repair label of Wm. Pelham, Cold Spring, N. Y.
Height 15 in. $150
Rippled Marine Timepiece, Seth Thomas Clock Co., Thomaston, Connecticut. Rosewood veneered case, painted dial, 30-hour time balance wheel movement.
Height 10 3/4 in. $150
Richard A. Bourne Co., Inc.

The first clock on the left is a Miniature Mahogany Veneer Ogee Clock, E. N. Welch, Forestville, Connecticut, c. 1860. It has a painted zinc dial, polychrome transfer door tablet "Buckingham Palace," housing a brass 30-hour time, strike-and-alarm movement. Paint loss on tablet.
Height 18 7/8 in. $325
The next clock is a Mahogany and Mahogany Veneer Steeple Clock, J. C. Brown, Forestville, Connecticut, c. 1845, with painted zinc dial, rippled molded door frame, geometric tablet, housing a 30-hour double frieze movement with auxiliary alarm mechanism. Minor damage.

Height 20 in. $575
Third in line is a Walnut Gingerbread Clock by the Ansonia Clock Company, Ansonia, Connecticut, c. 1870. This is the "Triumph" model.
Height 24 1/2 in. $350
On the right is a Walnut Calendar Clock, Ithaca Calendar Clock Company, Ithaca, New York, c. 1870. This is a No. 8 Library Shelf Model No. 106, housing an 8-day time-and-strike spring driven movement.
Height 24 3/4 in. $950
Robert W. Skinner, Inc.

Three scarce clocks, from left to right:

Scarce Shelf Clock, E. C. Brewster & Co., Bristol, Connecticut, c. 1845. An appealing clock constructed with choice rosewood veneers, a mirror tablet and upper tablet with black and gold border, mixed metal dial with Floral corner decorations, paper label, 30-hour time-and-strike movement with cast iron back plate and pendulum.

Height 24 in. $650

Scarce Steeple-on-Steeple Clock, E. N. Welch, Forestville, Connecticut, c. 1860. Rosewood veneered case with two exceptional tablets, enameled zinc dial, paper label, 8-day time, strike-and-alarm movement with pendulum.

Height 23 1/2 in. $750

Four Column Shelf clock, Seth Thomas, Plymouth Hollow, Connecticut, c. 1865. Mahogany case with four turned columns, exceptional painted tablet, paper label, enameled zinc dial, 30-hour time-and-strike movement with iron weights.

Height 26 in. $600

Richard A. Bourne Co., Inc.

The clock on the left is a Painted Cast Iron Front Mantel Clock, New Haven Clock Company, New Haven, Connecticut, c. 1860. Case with American eagle, shield, scales of justice, liberty cap and scrolls, painted metal dial with brass bezel, housing a brass 30-hour time-and-strike movement.

Height 13 in. $325

The middle clock is a Mahogany Veneer Cottage Clock, Bradley Manufacturing Company, Southington, Connecticut, c. 1860. Painted metal dial, reverse painted door tablet, housing a brass 8-day time-and-strike movement.

Height 15 3/8 in. $125

On the right is a Miniature Rosewood Veneer Ogee Shelf Clock, Seth Thomas, Thomaston, Connecticut, c. 1860. Painted metal dial, polychrome floral transfer tablet, housing a brass 8-day time-and-strike movement.

Height 15 3/4 in. $300

Robert W. Skinner, Inc.

Three different shelf clocks, from left to right:

Cottage Clock, Ansonia Clock Co., Ansonia, Connecticut, c. 1870. Original glass, rosewood veneered case, 8-day time-and-strike brass movement.

Height 18 1/4 in. $150

Iron Front Shelf Clock, Terhune & Edwards, c. 1850. Ornate, original condition, 30-hour time-and-strike brass movement.

Height 17 1/2 in. $165

Beehive Clock, E. N. Welch Manufacturing Co., Bristol, Connecticut, c. 1875. Rosewood veneered case, 8-day time, strike-and-alarm brass movement.

Height 18 1/2 in. $150

Richard A. Bourne Co., Inc.

Three exceptional examples, from left to right:
Scarce Miniature Clock Under Dome, Terryville Clock Co., Waterbury, Connecticut, c. 1875. Miniature skeleton clock under dome with a porcelain dial, pressed brass ornamentation including the maker's name, 8-day double wind time movement. The clock rests upon a base with painted decorations.
Height (excluding dome) 9 in. $450
Rare Miniature Shelf Clock, Silas B. Terry & Co., Terryville, Connecticut, c. 1852. Exceptional rosewood veneered case with gilded moldings and maker's name in gold on the base, painted zinc dial, 30-hour movement with time controlled by torsion twist of a flat steel rod on which are fastened 2 brass balls.
Height 8 1/2 in. $425
Scarce Papier-mâché Shelf Clock, Jerome Manufacturing Co., New Haven, Connecticut, c. 1850. Delicate gold decorations on the papier-mâché front and wood base, porcelain dial, 30-hour time movement with unusual Botdford Improved lever escapement. Paper label on back of papier-mâché facing.
Height 9 1/2 in. $350
Richard A. Bourne Co., Inc.

Three different shelf clocks, from left to right:
Very Rare Novelty Clock, movement: H. J. Davies, New York City; illuminating alarm mechanism: B. Bradley & Co., Boston, Massachusetts, c. 1865. A very interesting and rare clock with a mechanism on top that ignites a small lamp at the moment the alarm strikes. Solid walnut case with carved decoration, paper on zinc dial, paper label; retains what appear to be the original brass lamp and pendulum. 30-hour time-and-strike movement.
Height 14 1/2 in. $1500
Scarce Iron Front Shelf Clock, N. Pomeroy, Bristol, Connecticut, c. 1865. Cast iron case with mother-of-pearl inlay, gold decorations, painted zinc dial, 30-hour lever movement inscribed by the maker.
Height 10 1/4 in. $425
Miniature Cottage Clock, Seth Thomas Clock Co., Plymouth Hollow, Connecticut, c. 1860. Rosewood veneered case with a black and gold eglomise panel, painted zinc dial, paper label, 30-hour time movement.
Height 9 in. $250
Richard A. Bourne Co., Inc.

Three different shelf clocks, from left to right:
Rare Shelf Clock, S. B. Terry, Terryville, Connecticut, c. 1860. Nicely molded rosewood veneered case with a most unusual time-and-strike 30-hour movement; the movement utilizes a very odd placement of the gears and is mounted on a board. Opening the lower door reveals the maker's label.
Height 13 in. $950
Iron Front Shelf Clock, label: Forestville Manufacturing Co. ; movement: Brewster & Ingraham, Bristol, Connecticut, c. 1850. Although the name on the movement and label do not agree, careful inspection of both indicates that they are almost certainly original to each other. The iron case has fine gold decoration and is decorated with mother-of-pearl inlay; 8-day time-and-strike movement.
Height 14 in. $300
Flat Top Shelf Clock, Elisha Manross, Bristol, Connecticut, c. 1850. Rosewood veneered case with excellent moldings and quality cabinetry, bull's-eye in the bottom for pendulum visibility, painted zinc dial, 30-hour time-and-strike movement.
Height 14 1/4 in. $200
Richard A. Bourne Co., Inc.

Three different cast iron shelf clocks, from left to right:

Cast iron Shelf Clock, Terry & Andrews, Bristol, Connecticut, c. 1850. Mother-of-pearl inlay and painted decoration, painted zinc dial, maker's label, 8-day time-and-strike lyre movement, motion of the pendulum is visible through the small opening in the bottom.
Height 14 in. $375

Rare Cast Iron Shelf Clock, Upson Bros., Marion, Ohio, c. 1850. Gothic case with mother-of-pearl inlay and painted decoration, painted zinc dial, maker's label, brass time-and-strike movement marked "Morse #1. " The movement was probably purchased from Miles Morse & Co., Plymouth Hollow, Connecticut by Upson Bros.
Height 16 1/4 in. $400

Rare Large Round Top Iron Shelf Clock, Terry Clock Co., Waterbury, Connecticut, Pat. Dec 1, 1868. The largest of the series of this model, this clock features a calendar and an 8-day time movement; the case has scrolled decorations, and on the back is the maker's label.
Height 11 in. $800

Richard A. Bourne Co., Inc.

Three ornate shelf clocks, from left to right:

Very Rare Miniature Cast Iron Front Shelf Clock, Silas B. Terry, Plymouth Hollow, Connecticut, c. 1852. This rare clock has a superb maker's label on the back, and features Terry's patent torsion balance movement. The iron front is in original condition with mother-of-pearl inlay and painted scroll decoration. "Pat'd Oct 5, 1852" on movement.
Height 10 1/2 in. $950

Rare Miniature Cast Iron Clock, T. D. R. & Co., Bristol, Connecticut, c. 1855-1860. Very fine case with mother-of-pearl inlay and painted decoration, painted zinc dial, brass 30-hour balance wheel movement marked by the maker "T. D. R. & Co., Bristol, Ct., U. S. A. " The maker's name is possibly unique and is not listed in any of the lists of known clockmakers.
Height 10 1/2 in. $300

Rare Miniature Iron Shelf Clock, J. C. Brown Clock Co., Bristol, Connecticut, c. 1855. With mother-of-pearl inlay and painted scroll decoration, signed on dial and label, 30-hour balance wheel movement marked "Noah Pomeroy Bristol. "
Height 8 1/2 in. $275

Richard A. Bourne Co., Inc.

Three lovely shelf clocks, from left to right:

China Clock, Royal Bonn case, movement by Ansonia Clock Co., Ansonia, Connecticut, c. 1890. Fine cream-colored "La Orne" case with floral decorations, porcelain dial inscribed by the maker, and an 8-day time-and-strike movement.
Height 11 in. $400

Shelf Clock, American Clock Co., New york, c. 1875. Very unusual large gilt cast iron case, paper label and an 8-day time-and-strike movement.
Height 19 in. $550

China Clock, Ansonia Clock Co., Ansonia, Connecticut, c. 1900. A green case inscribed "Rainbow," with floral decorations, paper on zinc dial and an 8-day time-and-strike movement.
Height 11 1/4 in. $400

Richard A. Bourne Co., Inc.

Five different mantel clocks, from top to bottom, left to right:
Ceramic House Clock, Terregina Clock Co., Boston, Massachusetts, c. 1870. Architectural ceramic case with fine original label, 30-hour time-only movement.
Height 8 3/4 in. $100
Metal Cased Alarm Clock, Seth Thomas Clock Co., Thomaston, Connecticut, c. 1910-1920. One-day time-and-alarm movement with second bit.
Height 10 1/4 in. $75
Small Mantel Clock, Waterbury Clock Co., Waterbury, Connecticut, c. 1890. Time-and-strike movement which features outside escapement and rack-and-snail.
Height 9 1/2 in. $200
Iron Case Mantel Clock, Terry & Andrews, Bristol, Connecticut, c. 1840. Mother-of-pearl inlaid case with gold decoration, 8-day time-and-strike lyre movement.
Height 13 3/4 in. $175
Iron Case Mantel Clock, unknown maker, c. 1870. Floral decorated case with paw feet.
Height 16 in. $200
Richard A. Bourne Co., Inc.

Assortment of four shelf clocks, from left to right:
"Venetian" Model Shelf Clock, E. Ingraham & Co., Bristol, Connecticut, c. 1865. Rosewood veneered case and a maple "figure 8" door with an excellent floral tablet, paper on zinc dial, maker's label, 8-day time-and-strike movement.
Height 18 1/4 in. $250
Figural Clock, case by N. Muller, New York; movement by Crosby & Vosburgh, New York, c. 1875. Cast iron case painted with bronze paint, paper on zinc dial, paper label, 8-day time-and-strike movement.
Height 17 in. $225

Mantel Clock, Seth Thomas Clock Co., Thomaston, Connecticut, c. 1890. Cast white metal case with mermaids and other nautical motifs, porcelain dial, 8-day time-and-strike movement.
Height 16 1/4 in. $250
Iron Shelf Clock, T. Sperry, New York, c. 1855. Ebonized case with numerous mother-of-pearl inlays, floral decorations, 8-day time-and-strike movement.
Height 16 in. $175
Richard A. Bourne Co., Inc.

Six fine examples, from left to right and top to bottom:
Scarce Cottage Clock, New England Clock Co. Bristol, Connecticut, c. 1850. "Cigar Box" clock with an exceptional painted tablet, painted zinc dial inscribed by the maker, paper label, 30-hour time-and-alarm movement with pendulum.
Height 11 in. $600
Scarce Fusee Shelf Clock, Smith & Goodrich, Bristol, Connecticut, c. 1849. Reverse Ogee Shelf clock with mahogany veneer, exceptional painted tablet with eagle, shield and arrows, paper label, 30-hour time-and-strike fusee movement with pendulum.
Height 14 3/4 in. $400
Miniature Shelf Clock, Elisha Manross, Bristol, Connecticut, c. 1850. Unique rosewood veneered case with exceptional tablet depicting a balloon and American flags, paper label, painted zinc dial inscribed by the maker, 30-hour time-and-strike fusee movement with pendulum.
Height 20 in. $575
Round Gothic Shelf Clock, Terry & Andrews, Bristol, Connecticut, c. 1850. Fine mahogany veneered case with a painted tablet, paper label, painted zinc dial, 8-day time-and-strike lyre movement with pendulum.
Height 19 1/2 in. $700
Steeple Clock, J. C. Brown, Forestville, Connecticut, c. 1845. Veneered with vertical rosewood veneers having a fine etched glass tablet, painted zinc dial, paper label, 8-day time-and-strike movement with pendulum.
Height 19 1/2 in. $600
Richard A. Bourne Co., Inc.

Six representative shelf clocks, from top left to bottom right:
Shelf Clock, E. Ingraham & Co., Bristol, Connecticut, c. 1865. "Arch Column" with rosewood veneered case, blue and gold glass tablet, paper on zinc dial, paper label, 8-day time-and-strike movement with pendulum.
Height 17 in. $600
Scarce Shelf Clock, E. Ingraham & Co., Bristol, Connecticut, c. 1870. "Doric Mosaic Front" with black and gold tablet, paper on zinc dial, paper label, 8-day time-and-strike movement with pendulum.
Height 16 in. $300
Shelf Clock, E. Ingraham & Co., Bristol, Connecticut, C. 1860. "Arch Column" with rosewood veneered case, painted tablet, paper on zinc dial, paper label, 8-day time-and-strike movement with period pendulum.
Height 17 in. $300
Miniature Column and Cornice Shelf Clock, Seth Thomas Clock Co., Plymouth Hollow, Connecticut, c. 1860. Mahogany veneered case with gilded columns, painted tablet, painted zinc dial, paper label, 8-day lyre movement with period pendulum.
Height 19 in. $275
Empire Shelf Clock, Atkins Clock Co., Bristol, Connecticut, c. 1865. Very fine rosewood veneered case with two excellent black and gold tablets, painted zinc dial, paper label, 8-day time-and-strike movement with period pendulum.
Height 16 3/4 in. $550
Miniature Split Column Shelf Clock, Seth Thomas Clock Co., Plymouth Hollow, Connecticut, c. 1865. Mahogany veneered case with exceptional painted dial and glass, paper label, 30-hour time-and-strike movement with pendulum.
Height 19 in. $250
Richard A. Bourne Co., Inc.

A fine assortment of six clocks, featuring from the top left:
A Rare Rosewood Veneer Shelf Clock, Atkins Clock Manufacturing Company, Bristol, Connecticut, c. 1840. A fine painted tablet and a mirror tablet, enameled zinc dial, partial paper label, rare 30-day time-only double wind movement with lever spring.
Height 17 3/4 in. $1500
Next is a Shelf Clock, Jerome & Co., New Haven, Connecticut, c. 1875. Features "Gutta Percha" panels behind glass, rosewood veneered case, paper label 8-day time-and-strike movement with pendulum.
Height 16 in. $300
The third clock is a Rosewood Veneered Cottage Clock, unsigned, but Connecticut, c. 1875. Two black and gold tablets, painted zinc dial, 30-hour time-and-strike movement with pendulum.
Height 13 1/2 in. $150
On the left of the bottom row is a Scarce Steeple Clock, Brewster & Ingraham, Bristol, Connecticut, c. 1850. Features Charles Kirk cast iron back plate and lyre-shaped gong base, 8-day time-and-strike repeating movement marked by the maker, paper label, frosted tablet, enameled zinc dial stamped on the reverse by the maker, retains pendulum.
Height 20 in. $650
The middle clock is a Rosewood Veneered "Venetian" Model Shelf Clock, Welch Spring Co., Forestville, Connecticut, c. 1870. Gilded columns, black and gold tablet, paper label, 8-day time-and-strike movement stamped by E. N. Welch, retaining pendulum.
Height 19 in. $350
On the right is a lovely Mahogany Veneer Steeple Clock, Ansonia Brass & Clock Co., Ansonia, Connecticut, c. 1875. Decorated tablet, paper label, 8-day time-and-strike movement with pendulum.
Height 20 in. $450
Richard A. Bourne Co., Inc.

An assortment of shelf clocks, from left to right:
Venetian Figure "8" Shelf Clock, E. Ingraham Clock Co., Bristol, Connecticut, c. 1870. Mahogany case, fine old glass and original label; 8-day time-and-strike movement.
Height 15 3/4 in. $200
Empire Shelf Clock, E. Ingraham Clock Co., Bristol, Connecticut, c. 1850s. Rosewood case with rippled molded door, excellent early label which reads "Late Brewster & Ingraham. " 30-hour time, strike-and-alarm movement.
Height 17 1/2 in. $300
Late Miniature Pillar and Scroll Clock, Seth Thomas Clock Co., Thomaston, Connecticut, c. 1920. 8-day time-and-strike movement.
Height 17 in. $375
Small Empire Column Shelf Clock, Seth Thomas Clock Co., Thomaston, Connecticut, c. 1870. Rosewood case with original label, 8-day time-and-strike lyre movement, with "S. T. " hands.
Height 15 3/4 in. $$325
Cottage Clock, Seth Thomas Clock Co., Thomaston, Connecticut, c. 1860s. Rosewood veneered case with good original glass, 30-hour time-and-strike movement, half lyre movement, and patent dated dial.
Height 14 1/2 in. $200
Cottage Clock, New Haven Clock Co., New Haven, Connecticut, c. 1880. Rosewood veneered case, good original label, 8-day time-and-strike movement.
Height 13 3/4 in. $200
Richard A. Bourne Co., Inc.

Four shelf clocks, from left to right:
Twin Steeple Clock, Brewster & Ingraham, Bristol, Connecticut, c. 1840s. Rosewood veneered case, with frosted and cut door glass, original columns and finials.
Height 19 1/4 in. $550
Small Cottage Clock, Brewster & Ingraham, Bristol, Connecticut, c. 1845-1850. Mahogany case, 30-hour time-and-strike movement.
Height 13 3/4 in. $450
"Grecian" Model Shelf Clock, Ingraham Clock Co., Bristol, Connecticut, c. 1870. Rosewood case, with 30-hour time-and-strike movement, original graining on bezel.
Height 15 1/4 in. $500
"Doric" Model Shelf Clock, Ingraham Clock Co., Bristol, Connecticut, c. 1870. Very fine eagle decorated glass in black and gold, rosewood veneered case, 30-hour time-and-strike and alarm movement.
Height 16 1/4 in. $600
Richard A. Bourne Co., Inc.

Four styles of shelf clocks, from left to right:
Shelf Clock, E. Ingraham Clock Co., Bristol, Connecticut, c. 1875. "Doric" model, rosewood veneered case with gilded columns, gold tablet of a horse and rider, paper on zinc dial, paper label, 8-day time-and-strike movement.
Height 16 in. $225
Shelf Clock, E. Ingraham Clock Co., Bristol, Connecticut, c. 1878. "Doric" model with a rosewood veneered case, painted floral tablet, painted dial, maker's label and an 8-day time-and-strike movement inscribed by the maker.
Height 18 1/2 in. $200

Shelf Clock, E. N. Welch Manufacturing Co., Forestville, Connecticut, c. 1870. Rosewood veneered case with an exceptional black and gold tablet, paper label, and an 8-day time-and-strike movement with a "club tooth" escape wheel.
Height 18 1/2 in. $300
Rare Fusee Round Top Shelf Clock, Chauncey Jerome, Bristol, Connecticut, pat. 1839. Veneered with burl walnut, painted zinc dial inscribed by the maker, 8-day time-and-strike fusee movement.
Height 14 3/4 in. $600
Richard A. Bourne Co., Inc.

Very rare Dwarf Clock, probably Connecticut, case: c. 1800; movement: c. 1875. A diminutive cherry case with inlaid quarter fans and satinwood banding; case retains a fine patina, original hinges and delicate scrolled feet. An early label in the case interior reads, "Made by Silas Merriam, Bristol, Conn. " Movement and dial have been replaced by a spring-driven Connecticut movement, c. 1870.
Height 30 in. $2800
Richard A. Bourne Co., Inc.

Rare Painted Shelf Clock, New England, c. 1830. The rectangular case with shaped front panel of cut-out scrolled top with applied brass foliate devices above a circular opening framing a painted iron dial and brass weight driven movement, the waist with circular pendulum window joined with brass bezel on the rectangular base. The case retains original mahogany graining with simulated stringing and cross banding.
Height 28 1/8 in. $10,000
Robert W. Skinner, Inc.

A Cast Iron Front Mantel Clock, labeled Bradley & Hubbard, West Meriden, Connecticut, c. 1850. Polychrome, mother-of-pearl. and gilt decorated case, paper dial, housing an unsigned 8-day time-and-strike brass movement.
Height 20 in. $325
Robert W. Skinner, Inc.

Very Rare Shelf Clock, Miles Morse, Plymouth, Connecticut, c. 1849. Exceptional mahogany veneered case with inlaid satinwood panels on the cornice, delicate turned columns, exceptional enameled dial inscribed by Morse, painted tablet with mustard background, 8-day Salem Bridge movement with iron weights.
Height 26 in. $9500
Richard A. Bourne Co., Inc.

Three mantel clocks, from left to right:
Cast Iron Front Mantel Clock, Waterbury Clock Co., Waterbury, Connecticut, c. 1860. Beautiful cast iron case marked "Muller, N. Y.," with a 30-hour time-and-strike movement and pendulum opening in the lower part of the case.
Height 16 3/4 in. $225
Rare Cast Iron Novelty Clock, Ansonia Clock Co., New York, c. 1890. Gold painted case including busts of Beethoven and Wagner as well as other musical motifs. 30-hour balance wheel time movement with porcelain dial and a musical mechanism in the bottom of the case.
Height 10 3/4 in. $250
Double Column Mantel Clock, unsigned, c. 1910. Glass and white metal clock, 30-hour balance wheel movement.
Height 16 1/4 in. $200
Richard A. Bourne Co., Inc.

Two rare mantel clocks, from left to right:
Rare Double Dial Calendar Clock, "Aditi Model," E. N. Welch - movement, Forestville, Connecticut, c. 1875. Fine walnut case, two paper on zinc dials, the lower with unusual and rare indicators for the day and the month, Gale's patent perpetual calendar, 8-day time-and-strike movement.
Height 27 1/2 in. $1000
Rare Mantel Clock, Seth Thomas Clock Co., Thomaston, Connecticut, c. 1875. Solid walnut case with a painted gold tablet, original paper dial. Rare 8-day time-and-alarm movement that sounds with two hammers on a gong and bell.
Height 21 3/4 in. $500
Richard A. Bourne Co., Inc.

On the left we have a Victorian Mantel Clock, William L. Gilbert Clock Company, Winsted, Connecticut, c. 1896. This is a "Curfew" black clock, faux marble case, gilt metal bell and arch, scrolled feet, 8-day brass spring time-and-strike movement.
Height 17 1/2 in. $300
On the right is a desireable Crystal Palace Ansonia Mantel Clock, Ansonia, Connecticut, late 19th century. It has an 8-day time-and-strike movement, mirrored mahogany back, brass framed face, gilded figures of hunter and fisherman on oval walnut base, under glass dome.
Height (without dome) 16 5/8 in. $650
Robert W. Skinner Inc.

Five examples of mantel clocks:
The first on the left is a Mahogany Veneer Beehive Mantel Clock, Daniel Pratt & Sons, Reading, Massachusetts, c. 1860. There is a painted metal dial, polychrome transfer tablet in scroll and floral motif, 8-day brass spring driven time-and-alarm movement and oversized bell.
Height 18 3/4 in. $300
The second clock from the left is a Mahogany Veneer Beehive Mantel Clock by William Gilbert & Co., Winchester, Connecticut, c. 1866. There is a painted metal dial, polychrome transfer tablet with harbor scene. 8-day brass spring driven time-and-strike movement. Minor veneer damage.
Height 18 3/4 in. $200
Third is a Rosewood Mantel Clock, Welch Spring Company, Forestville, Connecticut, c. 1870. "Cary" model, housing a brass 8-day time-and-strike movement, paperweight type pendulum.
Height 20 1/2 in. $550
The fourth clock from the left is a Miniature Rosewood Veneer Ogee Mantel Clock, E. N. Welch Manufacturing Company, Forestville, Connecticut, c. 1850. It has a painted zinc dial, polychrome floral transfer tablet, housing a brass 8-day spring driven time, strike-and-alarm movement.
Height 18 1/4 in. $325
The clock on the right is a Mahogany "Patti" Mantel Clock, F. Kroeber, New York, c. 1874. This the Corinth model, with beveled glass tablet and sides, enamel dial, pendulum in the shape of a ship's wheel, 8-day brass spring driven time-and-strike movement.
Height 18 1/4 in. $950
Robert W. Skinner, Inc.

Four representative mantel clocks, from left to right:
Black Mantel Clock With Bell Top, William L. Gilbert Clock Co., Winsted, Connecticut, c. 1903. Marbleized wood case, 8-day time-and-strike, hour and half hour strike movement.
Height 18 1/2 in. $225
Iron Cased Mantel Clock, Kroeber, c. 1880. Ebonized finish, 8-day time-and-strike movement.
Height 12 1/2 in. $250
Fancy Victorian Mantel Clock, c. 1880s. Marked "New Haven" on the dial ring with a fine 8-day brass time-and-strike French movement. Decorative brass bell form ornament at top.
Height 23 1/2 in. $250
Adamantine Mantel Clock, Seth Thomas Clock Co., Thomaston, Connecticut, c. 1890. 8-day time-and-strike movement, original label.
Height 12 1/2 in. $295
Richard A. Bourne Co., Inc.

85.31.14

85.31.18

Black Flat Topped Mantel Clock, Seth Thomas Clock Company, Thomaston, Connecticut, c. 1880-1920. Three Doric columns on each side of face, lion's head with ring handles, leaf design feet, both of stamped brass, face is white with gold, hands and numbers are black, design on base and around face.
Height 11 1/8 in. $175
Museum of Connecticut History

Black Flat Topped Mantel Clock, New Haven Clock Company, New Haven, Connecticut, c. 1880-1920. Black grooved loop handles, black feet with leaf design, both of stamped brass, white face with black numbers and hands.
Height 11 3/8 in. $200
Museum of Connecticut History

85.31.9

85.31.4

Black Flat Topped Mantel Clock, Sessions Clock Company, Bristol, Connecticut, c. 1880-1920. Three gold and black Corinthian columns on each side of face, handles and feet are stamped brass, handles have a flower and leaf pattern, feet have three claws, face is white, hands are black.
Height 10 in. $150
Museum of Connecticut History

Black Flat-Topped Mantel Clock, Sessions Clock Company, Forestville, Connecticut, c. 1880-1920. Marbleized wood above and below three Corinthian columns, stamped brass feet and handles, both having a floral design, white face with gold, black Roman numerals, stenciled design around base and face.
Height 11 3/8 in. $200
Museum of Connecticut History

Flat Topped Mantel Clock, Sessions Clock Company, Bristol, Connecticut, c. 1880-1920. Yellowish marbleized wood above and below two Corinthian columns, brass decoration between the two columns, handles and feet are stamped brass, the handles are a lion's head with ring, the feet are leaf pattern, face is white with brass, black hands and numbers.
Height 10 1/4 in. $150
Museum of Connecticut History

Black Flat-Topped, Mantel Clock, E. Ingraham Company, Bristol, Connecticut, c. 1880-1920. Marbleized wood above and below three Corinthian columns on each side of face, handles and feet stamped brass, feet having a leaf and floral pattern, face is white with gold, hands are black.
Height 10 5/8 in. $200
Museum of Connecticut History

Black Flat Topped Mantel Clock, William L. Gilbert Clock Company, Winsted. Connecticut, c. 1880-1920. Rough marbleized wood above and below two green columns on either side of face, columns gold decorated, no handles, feet of stamped brass, white face, black hands and numbers.
Height 10 3/4 in. $150
Museum of Connecticut History

Black Flat Topped Mantel Clock, Sessions Clock Company, Bristol, Connecticut, c. 1880-1920. Marbleized wood above and below two Corinthian columns on each side of face, lion's head handles with rings, feet and handles gold painted stamped brass, white face with green design, hands and numbers are black.
Height 10 1/4 in. $150
Museum of Connecticut History

Black Flat Topped Mantel Clock, E. Ingraham Company, Bristol, Connecticut, c. 1880-1920. Marbleized wood above and column on each side of face, lion's head handles with ring, feet with leaf design, both of stamped brass painted black, white face with gold design, hands are black, numbers are gold.
Height 10 3/4 in. $150
Museum of Connecticut History

Black Flat Topped Mantel Clock, Seth Thomas Clock Company, Thomaston, Connecticut, c. 1880-1920. Brown marbleized wood on front, a Corinthian column with a face on it on either side of clock face, lion's head with ring handles, feet with curlicue design, both of stamped brass, gold colored face, hands and numbers black.
Height 11 in. $140
Museum of Connecticut History

Dome Top Mantel Clock, Sessions Clock Company, Bristol, Connecticut, c. 1880-1920. Green marbleized wood above and below three Corinthian columns on each side of face, handles and feet stamped brass with floral and leaf design, white face with gold, black hands and Roman numerals, design on dome, face and around base.
Height 12 7/8 in. $150
Museum of Connecticut History

Black Flat Topped Mantel Clock, William L. Gilbert Clock Company, Winsted, Connecticut, c. 1880-1920. Yellow marbleized wood above and below one wide column on either side of face, handles have a dot design, feet have a leaf design, both of stamped brass, white face, hands and numbers are black, design on base and around face.
Height 10 5/8 in. $165
Museum of Connecticut History

Black Eastlake Style Mantel Clock, E. Ingraham Company, Bristol, Connecticut, c. 1880-1920. Top rising to a peak, with ribbon design around, gold design in center of face, as well as edging and door rim, loop handles, leaf design feet, all trim in stamped brass.
Height 11 5/8 in. $200
Museum of Connecticut History

Black Ogee Top Mantel Clock, E. Ingraham Company, Bristol, Connecticut, c. 1880-1920. Red and black marbleized wood above and below columns on each side of face, lion's head handles with rings, claw feet, both of stamped brass painted gold, gold face, hands and numbers are black. On back is "Eight day half hour strike cathedral gong and patent regulator. "
Height 12 3/4 in. $150
Museum of Connecticut History

Black Dome Topped Mantel Clock, William L. Gilbert Clock Company, Winsted, Connecticut, c. 1880-1920. Marbleized wood above three Corinthian columns either side of face, leaf and floral design handles, leaf design feet, both of gold painted stamped brass, white face with black hands and numbers. Letter "G" on face.
Height 12 5/8 in. $150
Museum of Connecticut History

Black Curved Top Mantel Clock, William L. Gilbert, Winsted, Connecticut, c. 1880-1920. Three yellow marbleized columns with metal designs on each side of face, red marbleized wood with arch designs above the columns, black open loop handles, feet with leaf design, both of stamped brass, white face with letter "G," black hands and numbers.
Height 10 7/8 in. $150
Museum of Connecticut History

Chapter Two

Wall Clocks

Box Regulator #58 Clock, E. Howard & Co., Boston, Massachusetts, c. 1880. Cherry case, fine original condition, 8-day time-only weight driven movement.
Height 39 in. $2000
Richard A. Bourne Co., Inc.

Three different Regulator clocks, from left to right:
Box Regulator, unsigned, Connecticut, c. 1890. Molded solid oak with two decorated glass tablets, paper dial, and an 8-day time movement with period pendulum.
Height 38 in. $300
Long Drop Calendar Regulator, New Haven Clock Co., New Haven, Connecticut, c. 1880. "Bank" Model of solid oak with pressed designs, lower glass tablet with gold decorations, paper on zinc dial, 8-day time movement fitted with a simple calendar mechanism and large period pendulum. Original label on back of case.
Height 33 1/2 in. $450
Box Regulator, William Gilbert, Winsted, Connecticut, c. 1890. "Observatory" Model of solid oak with pressed designs, black and gold decorated upper tablet, paper on zinc dial, paper label, 8-day time movement with period pendulum.
Height 37 in. $375
Richard A. Bourne Co., Inc.

Three fine regulator clocks, from left to right:
#2 Regulator, Seth Thomas Clock Co., Thomaston, Connecticut, c. 1900. Fine quality oak case with a molded bracket, paper label, painted zinc dial with seconds indicator and maker's name, 8-day weight driven movement with brass pendulum, deadbeat escapement and maintaining power.
Height 36 in. $1000
Wall Regulator, Waltham Clock Co., Waltham, Massachusetts, c. 1900. Carved walnut case with a fine quality movement with deadbeat escapement and maintaining power inscribed "Waltham Clock Co." Note: lacks pendulum and weight.
Height 38 in. $425
Scarce Wall Regulator, Chelsea Clock Co., Boston, Massachusetts, c. 1920. #2 Model, fine walnut case, painted zinc dial with seconds indicator inscribed by the maker, 8-day movement with deadbeat escapement and maintaining power, retains original pendulum.
Height 38 in. $1350
Richard A. Bourne Co., Inc.

Three regulator clocks, from left to right:
Regulator #2, Seth Thomas Clock Co., Thomaston, Connecticut, c. 1880. Mahogany and mahogany veneered case, painted zinc dial paper label, 8-day time weight driven movement.
Height 36 in. $850
Scarce Regulator, #3, Seth Thomas Clock Co., Thomaston, Connecticut, c. 1930. Solid oak case with painted zinc dial with seconds indicator, 8-day time movement with deadbeat escapement and maintaining power and marked "#3. " Retains pendulum and paper label.
Height 42 in. $1150
#2 Regulator, New Haven Clock Co., New Haven, Connecticut, c. 1900. Solid oak case, paper on zinc dial, 8-day time weight driven movement with pendulum.
Height 35 in. $800
Richard A. Bourne Co., Inc.

Three different examples of regulator clocks, from left to right:
Short Drop Regulator, Seth Thomas Clock Co., Thomaston, Connecticut, c. 1875. Rosewood veneered case with black and gold tablets, painted zinc dial, paper label and an 8-day time movement.
Height 24 1/2 in. $300
Long Drop Calendar Regulator, New Haven Clock Co., New Haven, Connecticut, c. 1890. Exceptional pressed oak case, "Regulator" tablet, paper on zinc dial with calendar, paper label, 8-day time movement.
Height 33 in. $400
Short Drop Regulator, Seth Thomas Clock Co., Thomaston, Connecticut, c. 1875. Walnut veneered case with black and gold tablet, painted zinc dial, paper label, and an 8-day time movement with pendulum.
Height 24 1/2 in. $275
Richard A. Bourne Co., Inc.

Three Short Drop Regulator clocks, from left to right:
Short Drop Regulator Clock, Daniel Pratt and Sons, Boston, Massachusetts, c. 1870. Rosewood veneered case; includes an over label of the Waterbury Clock Co., 8-day time-only movement.
Height 21 1/2 in. $295
Short Drop Regulator Clock, The New Haven Clock Co., New Haven, Connecticut, c. 1870-1880. Exceptional "Lucky Strike" label on front, black and gold lower glass signed "Haskell & Adams Boston, Mass. " 8-day time-only movement.
Height 24 in. $350
Short Drop Regulator Clock, Seth Thomas Clock Co., Thomaston, Connecticut, c. 1910. Solid walnut case, 8-day time-and-strike movement.
Height 17 1/2 in. $350
Richard A. Bourne Co., Inc.

Regulator #75 Clock, E. Howard & Co., Boston, Massachusetts, c. 1880. Cherry case, 8-day time-only weight driven movement. Height 32 in. $1350
Richard A. Bourne Co., Inc.

Three prime examples of Regulator wall clocks, from left to right:
Model #2 Regulator, Seth Thomas, Thomaston, Connecticut, c. 1890. Quarter grain oak veneered case, fine label, excellent finish, 8-day weight driven time-only movement with second bit.
Height 36 in. $950
Model #1 Regulator, Seth Thomas, Thomaston, Connecticut, c. 1860s. Wonderful original glass with eagle and star, "Regulator" in the bottom section, excellent label, 8-day time-only movement.
Height 34 in. $800
Model #2 Regulator, Seth Thomas, Thomaston, Connecticut, c. 1880s. Walnut veneered case, 8-day time-only movement with second bit.
Height 37 in. $875
Richard A. Bourne Co., Inc.

Regulator #70 Clock, E. Howard & Co., Boston, Massachusetts, c. 1880. Quarter grained oak case, 8-day time-only weight driven movement. Height 32 in. $1375
Richard A. Bourne Co., Inc.

Two representative Regulator clocks are, from the left, an Oak Regulator Timepiece, Waltham Clock Co., Waltham, Massachusetts, c. 1900, with a round wooden bezel, painted dial, housing a brass 8-day weight driven movement.
Length 32 1/2 in. $900
The right hand clock is a Mahogany Regulator Wall Clock, Ansonia Clock Co., Ansonia, Connecticut, c. 1880. "Queen Elizabeth," paper dial, housing an 8-day time-and-strike movement.
Length 37 in. $625
Robert W. Skinner, Inc.

Three lovely regulator clocks, from left to right:
Regulator Clock, Seth Thomas Co., Thomaston, Connecticut, c. 1910. Model #1, 8-day weight driven time-only movement with second bit.
Height 36 in. $1250
In the middle is a Regulator Wall Clock, Seth Thomas Co., Thomaston, Connecticut c. 1900. Model #30, mahogany case, 8-day time-only movement with second bit.
Height 50 in. $1600
On the right is a Waltham Regulator Clock, c. 1900-1910. Oak Case, with 8-day time-only movement.
Height 34 in. $650
Richard A. Bourne Co., Inc.

Model #1 Regulator Wall Clock, Seth Thomas Clock Co., c. 1860-1870. Mahogany case, wonderful old original label, 8-day time-only movement.
Height 34 1/4 in. $600
Richard A. Bourne Co., Inc.

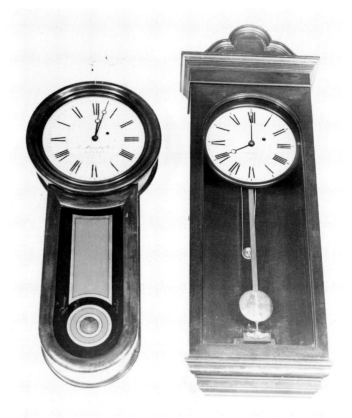

Two Howard Regulator clocks, from left to right:
Howard Keyhole Regulator Clock, Howard & Co., Boston, Massachusetts, c. 1880. Exceptional clock in fine original condition, fine grained case, exceptional fine signed dial in script with rare Howard label illustrating ten different Howard clocks, 8-day weight driven time-only movement.
Height 31 in. $7500
Howard #59 Box Regulator Clock, Howard & Co., Boston, Massachusetts, c. 1890s. Solid mahogany case, excellent finish with original instruction label, 8-day weight driven time-only movement.
Height 40 in. $2100
Richard A. Bourne Co., Inc.

Howard #70 Regulator Clock, Howard & Co., Boston, Massachusetts, c. 1880. Solid Circassian walnut facings with cherry side, script signature on face, 8-day time-only movement.
Height 41 1/2 in. $2100
Richard A. Bourne Co., Inc.

Two Regulator wall clocks, from left to right:
Model #1 Regulator Extra Clock, Seth Thomas, Thomaston, Connecticut, c. 1860. Rosewood veneered case, rare signed pendulum, superb glass and label, tulip hands, 8-day double wind time-and-strike movement with second bit.
Height 44 in. $2850
Regulator Wall Clock, Waltham Clock Company, Waltham, Massachusetts, c. 1900. Quarter grain solid oak case, 8-day weight driven time-only movement.
Height 37 1/2 in. $1000
Richard A. Bourne Co., Inc.

Scarce Wall Regulator, E. Howard Clock Co., Boston, Massachusetts, c. 1880. No. 11 Model, more commonly known as the "Keyhole." In a cherry case with traces of the grained decoration, a red, black and gold glass, painted zinc dial inscribed by the maker, and an 8-day time movement also inscribed by the maker, retaining the original cast iron weight.
Height 31 1/2 in. $2750
Richard A. Bourne Co., Inc.

Two fine Regulator clocks, from left to right:
Scarce Regulator Clock, Seth Thomas, Thomaston, Connecticut, c. 1870. "Early" Regulator #1 constructed of mahogany and mahogany veneers, with black and gold "Regulator" glass, enameled zinc dial, fine black and gold paper label, 8-day time movement with maintaining power, dead beat escapement, iron weight and pendulum.
Height 34 1/4 in. $695
Regulator Clock, Seth Thomas, Thomaston, Connecticut, c. 1880. Regulator #1 constructed of walnut and walnut veneers with drop finials, enameled zinc dial with seconds indicator inscribed by the maker, 8-day time movement with dead beat escapement, maintaining power, brass weight and nickel plated pendulum with engraved decoration. Paper label in bottom of case.
Height 36 1/2 in. $850
Richard A. Bourne Co., Inc.

Three fine regulator clocks, from left to right:
Very Rare Miniature Regulator #89, E. Howard Co., Boston, Massachusetts, c. 1880. Removed from the storeroom of the Howard Sales Office on Washington Street, Boston, Massachusetts, this clock has some very unusual and unique original features, including the paneled door and a small hinged door on the left side of the case. 8-day time movement marked by the maker and "111/2" with deadbeat escapement and stop works; painted zinc dial also inscribed by the maker.
Height 34 in. $900
#2 Regulator Clock, Seth Thomas Clock Co., Thomaston, Connecticut, c. 1875. Oak case with painted zinc dial with seconds indicator, 8-day time movement with maintaining power, deadbeat escapement, brass weight and pendulum.
Height 36 1/4 in. $875
Regulator Clock, E. Howard Clock Co., Boston, Massachusetts, c. 1885. Pine case with a Howard #70 movement and pendulum, painted zinc dial, period iron weight.
Height 33 in. $575
Richard A. Bourne Co., Inc.

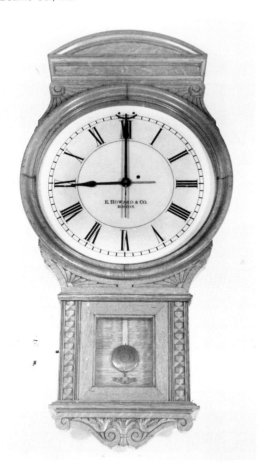

Howard Oak Regulator Clock, Howard & Co., Boston, Massachusetts, c. 1880. Solid quarter grain carved oak case, with #70 8-day weight driven time-only movement. Possibly a prototype, absolutely mint in all respects.
Height 43 1/2 in. $4700
Richard A. Bourne Co., Inc.

Rare Regulator, possibly by Welch Spring & Co., Bristol, Connecticut, c. 1880. Fine rosewood veneered case, having a laminated and turned bezel, painted zinc dial with seconds indicator, very unusual two weight 30-day upside-down movement, retaining two period brass weights and a fine brass pendulum with gilded rod.
Height 57 in. $1450
Richard A. Bourne Co., Inc.

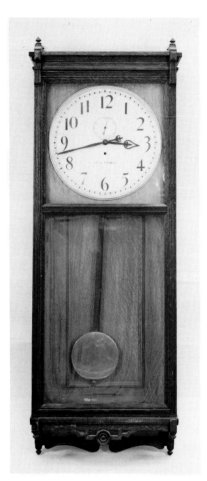

Wall Regulator, E. Howard & Co., Boston, Massachusetts, c. 1885. Similar to a No. 89 Model, but larger. Fine walnut case with turned finials and reeded moldings. Originally a watch clock, the components have been removed, which explains the holes in the dial. 8-day time movement marked "E. Howard & Co.," dead beat escapement and a fine nickel plated pendulum.
Height 72 1/2 in. $1050
Richard A. Bourne Co., Inc.

Scarce Regulator, Seth Thomas Clock Co., Thomaston, Connecticut, c. 1880. Model #31, fine oak case with turned finials and reeded moldings, painted zinc dial with seconds indicator, 8-day time movement with dead beat escapement retaining a period brass weight and nickel pendulum, movement marked with Seth Thomas trademark.
Height 69 in. $1100
Richard A. Bourne Co., Inc.

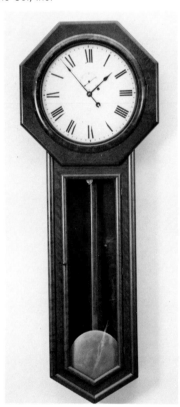

Rare Regulator Clock, Seth Thomas Clock Co., Thomaston, Connecticut, c. 1880. Model #18, with walnut and walnut veneered case, painted zinc dial with seconds indicator, Seth Thomas trademark, fine 8-day time movement with dead beat escapement, maintaining power, nickel plated pendulum with damascene decoration.
Height 54 in. $2000
Richard A. Bourne Co., Inc.

#72 Regulator, E. Howard & Co., Boston, Massachusetts, c. 1890. Carved and reeded solid oak case fitted with Howard #1 movement with Graham dead beat escapement, maintaining power, stop works, nickel plated pendulum with damascene decoration and the original iron weight inscribed #1, 14 inch diameter painted zinc dial inscribed by the maker with seconds indicator.
Height 63 1/2 in. $2100
Richard A. Bourne Co., Inc.

Two exceptional regulator clocks, from left to right:
Regulator Clock, E. Howard & Co., Boston, Massachusetts, c. 1895. Said to have been removed from the Howard Clock Co. factory in Roxbury in 1895, this clock features a fine walnut case, painted zinc dial, 8-day weight driven movement with damascene decorations on the plates and an unusual detached striking mechanism that strikes on a bell mounted on top, retains a fine large pendulum and iron weight.
Height 60 1/2 in. $1400
Scarce Regulator Clock, unsigned, probably by Atkins Clock Co., Connecticut, c. 1880. Rosewood veneered case, painted zinc dial with seconds indicator, an unusual two weight 8-day time movement with deadbeat escapement; retains brass pendulum bob and a gilded rod.
Height 52 in. $850
Richard A. Bourne Co., Inc.

Rare Regulator, S. B. Terry, Plymouth, Connecticut, c. 1835. Beautiful mahogany case with a turned wood bezel that opens to an excellent painted wood dial; a lower door with a crotch mahogany panel that opens to a paper label and a very unusual 8-day weight-driven time movement with a round front plate.
Height 34 1/2 in. $4800
Richard A. Bourne Co., Inc.

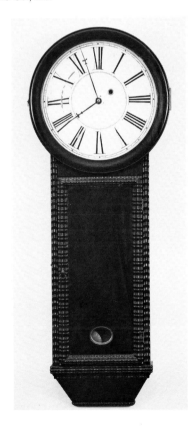

#41 Regulator Clock, E. Howard Clock Co., Boston, Massachusetts, c. 1875. Solid walnut case with carved crest and bracket, a door with a red and gold tablet that opens to the dial inscribed "E. Howard & Co., Boston. " Paper label on inside of door, 8-day time movement marked "E. Howard & Co., Boston. "
Height 48 1/2 in. $2850
Richard A. Bourne Co., Inc.

Rare Early Regulator, possibly A. Willard, Jr., Boston, Massachusetts, c. 1830. Choice mahogany case with applied rippled moldings, turned laminated bezel, painted zinc dial, quality 8-day banjo-type movement retaining an original lead weight, and having a square pendulum rod; this feature is associated with A. Willard Jr.
Height 34 1/2 in. $1800
Richard A. Bourne Co., Inc.

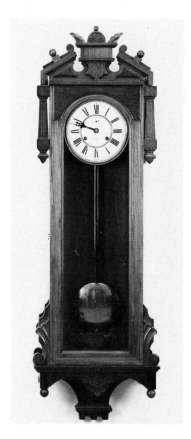

Regulator Clock, Ansonia Clock co., Ansonia, Connecticut, c. 1880. Solid walnut case with carved and reeded decorations, porcelain dial with seconds bit and maker's trademark, large brass pendulum and regulator gauge, and an unusual two-weight time movement that is probably a 30-day, also retaining two period brass weights.
Height 49 1/4 in. $1250
Richard A. Bourne Co., Inc.

Electric Wall Regulator Clock, New York Standard Watch Co., New York, c. 1900. In a superb solid cherry case with a very simple battery movement; batteries are stored in the bottom bracket, painted zinc dial and a beautiful brass pendulum.
Height 80 1/2 in. $1050
Richard A. Bourne Co., Inc.

Long Case Regulator Clock, Seth Thomas, Thomaston, Connecticut, c. 1900. Solid quarter grain oak case, bank model with 30-hour time-only movement with second bit.
Height 68 in. $1275
Richard A. Bourne Co., Inc.

Absolutely mint Howard Regulator #60, Howard & Co., Boston, Massachusetts, c. 1880. Superb black walnut case, mercury pendulum; 8-day dead beat escapement movement with second bit.
Height 6 ft. 8 in. $11,500
Richard A. Bourne Co., Inc.

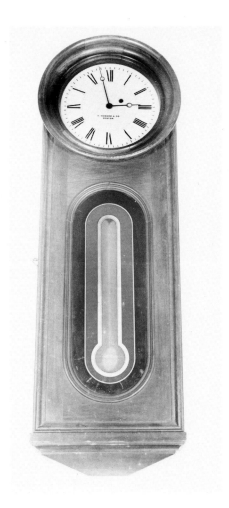

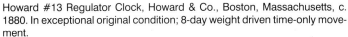

Howard #13 Regulator Clock, Howard & Co., Boston, Massachusetts, c. 1880. In exceptional original condition; 8-day weight driven time-only movement.
Height 42 in. $3300
Richard A. Bourne Co., Inc.

A nice example of a Walnut Regulator Clock, Ansonia Clock Co., Ansonia, Connecticut, c. 1880. This is the "Capitol" model, arched crest with cast metal portrait bust, paper dial, silver gilt decorated door glass, housing a brass 8-day time-and-strike movement.
Length 49 1/2 in. $700
Robert W. Skinner, Inc.

Large Walnut Wall Regulator, E. Howard & Company, Boston, Massachusetts, c. 1860. No. 7, painted dial, weight driven movement, stamp "E. Howard & Co., Boston" in Old English lettering.
Length 63 1/2 in. $2200
Robert W. Skinner, Inc.

Rare Federal Gilt Gesso Painted Banjo Timepiece, Aaron Willard, Boston, Massachusetts, c. 1808. The 8-day weight driven movement stamped "A. Willard Boston 1808," eglomise rectangular tablet depicting "Hercules assisting Atlas in supporting the Globe," signed on reverse "A. Willard Jr. and Spencer Nolen. Washington Street, Boston. "
Height 45 in. $225,000
Robert W. Skinner, Inc.

Banjo Clock, Horace Tift, Attleboro, Massachusetts, c. 1820. Mahogany case with gilded bracket and two frames, two excellent professionally painted tablets, painted iron dial inscribed "Tift," surmounted by an eagle finial; 8-day time movement retaining a period lead weight.
Height 37 in. $1450
Richard A. Bourne Co., Inc.

Banjo Clock, unknown maker, Massachusetts or Rhode Island, c. 1825. Howard-type poplar case with a gilded bracket, gilded frames and bezel, two excellent black and gold tablets, 8-day time movement retaining a period weight.
Height (excluding finial) 33 1/2 in. $700
Richard A. Bourne Co., Inc.

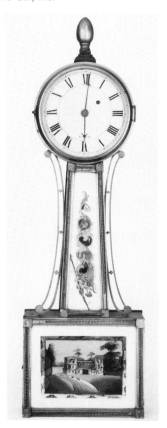

Banjo Clock, Roxbury School, c. 1810. Mahogany case with cross banded frames and painted glass tablets, painted convex dial with Arabic numerals, lower glass is Coat of Arms, 41/16-inch 8-day time movement with step train and T-bridge suspension.
Height (excluding finial) 29 3/4 in. $800
Richard A. Bourne Co., Inc.

Rare Banjo Clock, J. Seward, Boston, Massachusetts, c. 1835. Fine gold front banjo, with painted tablets, brass side arms and bezel, painted iron dial inscribed "Seward," acorn finial, and an 8-day time movement with original weight.
Height (excl. finial) 29 1/2 in. $2100
Richard A. Bourne Co., Inc.

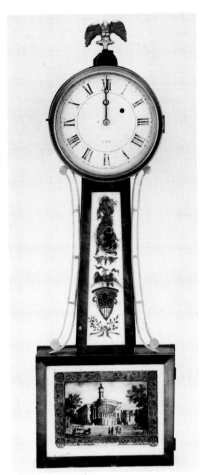

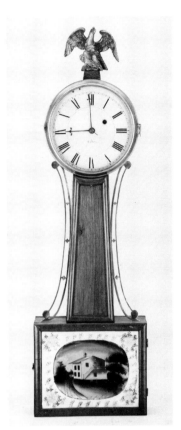

Banjo Clock, unsigned, Massachusetts, c. 1820. Handsome mahogany case with two fine painted tablets, the lower of "Girard's Bank," a painted iron dial inscribed "Zacheus Gates," brass side arms, bezel and eagle finial. 8-day time movement with a period cast iron weight.
Height 32 in. $1350
Richard A. Bourne Co., Inc.

Banjo Clock, unsigned, Massachusetts, c. 1820. Mahogany case with a fine painted tablet, mahogany throat panel, painted iron dial inscribed "Zacheus Gates," brass eagle finial and side arms, and an 8-day banjo-type movement.
Height 33 1/4 in. $750
Richard A. Bourne Co., Inc.

Presentation Banjo Clock, John Sawin, Boston, Massachusetts, c. 1815. A beautiful clock with a gilded bracket and frames, two beautifully painted tablets, painted dial, eagle finial, and an 8-day weight driven movement retaining a period weight.
Height (including finial) 41 1/2 in. $1500
Richard A. Bourne Co., Inc.

Federal Mahogany Inlaid Banjo Timepiece, Simon Willard, Roxbury, Massachusetts, c. 1820. The mahogany case with circular molded brass bezel and painted iron dial above the square tapering throat eglomise tablet of oak leaves and acorns contained by a teal-green vine border, on a rectangular tablet below with a demilune and oak leaf motif inscribed "S. Willards Patent," contained in inlaid flames with stringing and crossbanding flanked by pierced brass side arms.
Note: The interior tin plate inscribed "E. Taber, Roxbury" for Elnathan Taber, Willard's "favorite apprentice." The case is stamped "J. Davis."
Height 33 in. $28,500
Robert W. Skinner, Inc.

A Mahogany and Crossbanded Veneer Banjo Timepiece, Waltham Clock Co., Waltham, Massachusetts, c. 20th century. A cast brass eagle finial, brass bezel and side arms, painted metal dial, reverse painted throat and door tablets "Boston State House," bracket with balls on acorn finials, housing an 8-day brass weight driven movement, signed and numbered 7350.
Length 42 in. $1900
Robert W. Skinner, Inc.

Waterbury Presentation Banjo Timepiece, Waterbury Clock Co., Waterbury, Connecticut, c. 20th century. Ball finial, brass bezel, porcelain enamel dial, reverse painted throat and door tablet in beaded gilt frames, bracket with gilt balls and acorn pendant, housing a brass 8-day weight driven movement.
Height 44 in. $1250
Robert W. Skinner, Inc.
Willard-type Banjo Clock, c. 1820. T-bridge weight driven 8-day time-only movement, door and throat glass original to clock, but flaking, original weight and eagle finial.
Height 34 in. $1500
Richard A. Bourne Co., Inc.

Banjo Clock, New Haven Clock Co., New Haven, Connecticut, c. 1920. Mahogany cased, 8-day time-and-strike spring-driven movement.
Height 36 1/2 in. $375
Octagon Face Banjo Clock, New Haven Clock Co., New Haven, Connecticut, c. 1920. Mahogany, with triple wind time-and-strike 8-day spring-driven movement, quarter hour strike.
Height 41 1/2 in. $550
Richard A. Bourne Co., Inc.

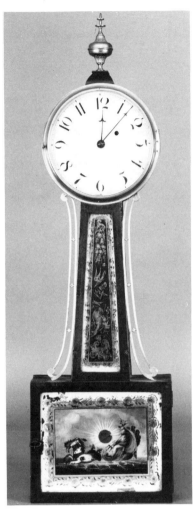

#4 Banjo Clock, Howard & Davis, Boston, Massachusetts, c. 1860. Case shows excellent graining, 8-day time-only weight driven brass movement.
Height 32 in. $700
Richard A. Bourne Co., Inc.

Rare Banjo Clock, William Cummens, Boston, Massachusetts, c. 1810. A beautiful finish to this fine old banjo clock, with cross banded mahogany frames, exceptional painted tablets, original brass finial and a high quality 8-day T-bridge movement.
Height (with finial) 34 in. $3500
Richard A. Bourne Co., Inc.

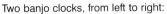

A Very Fine Presentation Banjo Clock, Elmer O. Stennes after Simon Willard. Case marked "10/74," 8-day weight-driven time-only movement.
Height 41 3/4 in. $2300
Richard A. Bourne Co., Inc.

Two banjo clocks, from left to right:
Howard & Davis #2 Banjo Clock, Howard & Davis Co., Roxbury, Massachusetts, c. 1860. Fine grained case, excellent glasses, made for "Riggs, Philadelphia," and so marked on the works, 8-day weight driven time-only movement.
Height 43 3/4 in. $2450
Howard & Davis #3 Banjo Clock, Howard & Davis Co., Roxbury, Massachusetts, c. 1860. Fine graining, excellent glasses and dial, 8-day weight driven time-only movement.
Height 38 in. $2900
Richard A. Bourne Co., Inc.

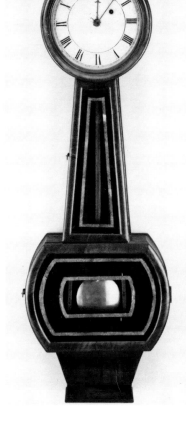

Howard-Type Banjo Clock, unsigned, Massachusetts, c. 1840. Softwood case, two black and gold painted tablets, painted zinc dial, turned wood bezel, 8-day time movement retaining an old weight and pendulum.
Height 29 in. $1100
Richard A. Bourne Co., Inc.

Early Howard-Type Banjo Clock, unsigned, Massachusetts, c. 1820. Fine mahogany case with a very unusual and unique drop on the bottom, two black and gold glasses in hinged frames, turned wood finial and bezel, 8-day time movement with a period lead weight and pendulum.
Height 36 1/2 in. $1250
Richard A. Bourne Co., Inc.

Baltimore Banjo Clock, maker unknown, c. 1870. Fine mahogany case, marked on the dial, "W. H. C. Riggs, Philada," Howard glasses, 8-day weight driven time-only movement.
Height 33 1/2 in. $3000
Richard A. Bourne Co., Inc.

Walnut Figure Eight Banjo Timepiece, E. Howard & Co., Boston, Massachusetts, c. 1857. This is a No. 8, painted metal dial, black and gold throat tablet red, black and gold door tablet, housing an 8-day weight driven movement.
Length 44 in. $7000
Robert W. Skinner, Inc.

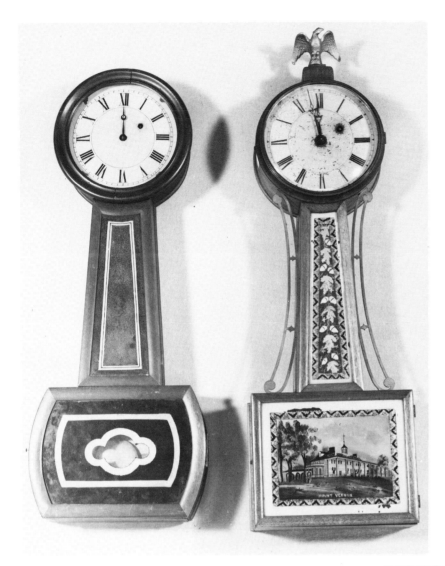

Two Banjo clocks, from left to right:
Banjo Clock, unsigned, Massachusetts, c. 1870. A Howard-type Banjo clock of walnut with black and gold glasses, painted zinc dial, 8-day time movement with iron weight and pendulum.
Height 28 3/4 in. $850
Rare Banjo Clock, Little & Eastman, Boston, Massachusetts, c. 1900. Mahogany with brass sidearms, bezel and eagle finial, two painted tablets, the lower depicting Mount Vernon, the upper in floral decoration, painted zinc dial, 8-day time movement with lead weight and pendulum; movement stamped "Little & Eastman, Boston. "
Height 32 in. $900
Richard A. Bourne Co., Inc.

Two extremely fine banjo clocks, from left to right:
Banjo Clock, unsigned, probably Massachusetts, c. 1830. Mahogany case with fine inlay, painted iron dial inscribed "O. Brokaw. . . (?)," two fine eglomise tablets, turned finial and an 8-day time movement.
Height 40 1/2 in. $1850
Presentation Banjo Clock, unsigned, probably Massachusetts, c. 1830. Mahogany case with gilded frames and bracket, two eglomise tablets, the lower depicting the Boston State House, and a rare 8-day time-and-alarm movement with an old iron weight.
Height 37 in. $1300
Richard A. Bourne Co., Inc.

Three exceptional banjo clocks, from left to right:
Banjo Clock, unsigned, Massachusetts, c. 1880. Walnut case with two black and gold tablets, painted zinc dial, 8-day time weight driven movement with iron weight and pendulum.
Height 29 in. $850
Banjo Clock, Howard & Davis, Boston, Massachusetts, c. 1850. Grain-painted poplar case with two black and gold tablets open in the centers to reveal the gilded pendulum rod and brass bob, paper on zinc dial inscribed by the maker, quality 8-day time movement inscribed "Howard & Davis, Boston. " With iron weight and original pendulum.
Height 32 in. $1100
Banjo Clock, Horace Tift, North Attleboro, Massachusetts, c. 1840. Fine mahogany and mahogany veneered case with two black and gold tablets, carved side arms, painted zinc dial inscribed "H. Tift," 8-day time movement with original weight and pendulum. A special feature is the original weight with the initials "H. T. " cast into it; also retains the original finials and plinth.
Height 33 1/4 in. $950
Richard A. Bourne Co., Inc.

Three exceptional banjo clocks, from left to right:
Presentation Banjo Clock, Waltham Clock Co., Waltham, Massachusetts, c. 1930. Excellent quality throughout with a mahogany case, two painted tablets of George Washington and Mount Vernon, engraved painted dial, gilded finials top and bottom, 8-day time weight driven movement marked by the maker, with deadbeat escapement, maintaining power, stop works and period weight and pendulum.
Height 41 in. $1550
Banjo Clock, Horace Tift, Attleboro, Massachusetts, c. 1840. An elegant example with gilded frames and finial, superb painted tablets of the naval engagement "Chesapeake and the Shannon," painted iron dial inscribed "Tift," and an 8-day weight driven movement.
Height 33 1/4 in. $1100
Banjo Clock, unsigned, Massachusetts, c. 1840. Mahogany case with two exceptional tablets, the lower depicting a naval engagement on Lake Erie, 8-day time movement with period lead weight and pendulum, painted iron dial.
Height 33 in. $1150
Richard A. Bourne Co., Inc.

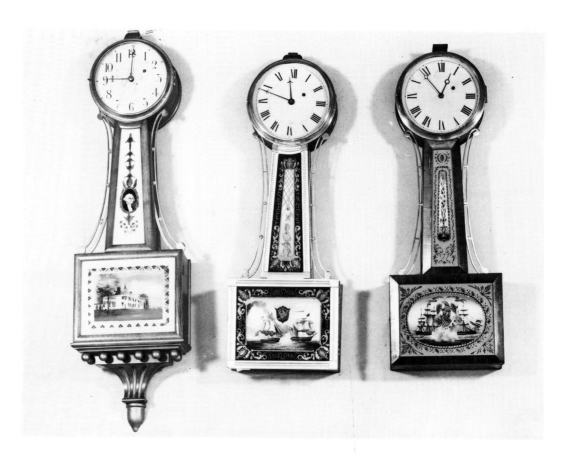

Three choice banjo style clocks, from left to right:
Banjo Clock, unsigned, Massachusetts, c. 1830. Mahogany front case with applied rope and gilding, gold front with a nautical scene on the lower glass, painted zinc dial, 8-day time movement.
Height 33 1/2 in. $600
Banjo Clock, unsigned, Massachusetts, c. 1840. Grained poplar case with two black and gold tablets, painted zinc dial, 8-day time movement retaining period iron weight.
Height 29 in. $950
Banjo Clock, unsigned, Massachusetts, c. 1830. Originally a mahogany front case, rope was applied to the frames and then gilded, gold front with nicely painted tablets, painted iron dial, 8-day time movement attached with fillister head mounting bolts; retains original iron weight.
Height 33 in. $1000
Richard A. Bourne Co., Inc.

Four representative banjo clocks, from left to right:
Miniature Banjo Clock, Waltham Watch Co., Waltham, Massachusetts, c. 1930. Solid mahogany case with carved bracket, brass case hardware and two exceptional painted tablets of George Washington and Mount Vernon. High quality 8-day jeweled lever movement, painted dial.
Height 21 in. $800
Miniature Banjo Clock, New Haven Clock Co., New Haven, Connecticut, c. 1920. Mahoganized soft wood case with two decorated tablets, silvered dial and 12-day lever movement.
Height 22 in. $300
Small Banjo Clock, Seth Thomas Clock Co., Thomaston, Connecticut, c. 1920. Mahoganized soft wood case, with brass hardware, two decorated Nautical tablets, engraved silver dial and an 8-day time-and-strike lever movement.
Height 26 in. $300
Banjo Clock, Sessions Clock Co., Forestville, Connecticut, c. 1920. "Revere" Model with mahoganized soft wood case, engraved silver dial, and two decorated tablets of George Washington and Mount Vernon; 8-day time-and-strike movement with hour strike on two gongs, retains paper label.
Height 31 1/2 in. $375
Richard A. Bourne Co., Inc.

Four different banjo clocks, from left to right:
Miniature Banjo Clock, New Haven Clock Co., New Haven, Connecticut, c. 1920. Mahogany case with two decorated tablets, the lower depicting Mount Vernon, porcelain dial, 8-day lever movement.
Height 15 1/2 in. $200
Medium Banjo Clock, New Haven Clock Co., New Haven, Connecticut, c, 1920. Exceptional, with decorated tablets depicting Mount Vernon, 3-train movement striking Westminster chimes on the quarter hours.
Height 25 in. $500
Medium Banjo Clock, New Haven Clock Co., New Haven, Connecticut, c.

1920. Mahogany case with decorated tablets, the lower one depicting Mount Vernon; silvered dial with raised numerals, 8-day time-and-strike pendulum movement.
Height 25 in. $250
Miniature Banjo Clock, Seth Thomas Clock Co., Thomaston, Connecticut, c. 1920. Soft wood case with decorated Nautical tablets, engraved silvered dial, 30-hour 4 jewel lever time movement.
Height 19 in. $100
Richard A. Bourne Co., Inc.

From left to right, we have a Mahogany Banjo Timepiece, Massachusetts, c. 1840. It has an acorn finial, painted iron dial, mahogany throat tablet, reverse painted door tablet, housing a brass 8-day A-shaped weight driven movement.
Length 32 1/2 in. $800
The next is a Mahogany Banjo Timepiece, Eastern United States, c. 1830. A round wooden bezel, painted metal dial, wooden side arms, black and gold reverse painted throat and tablet, housing a brass 8-day weight driven movement.
Length 33 in. $800
The third clock is a Mahogany Banjo Timepiece, Seth Thomas Clock Company, Thomaston, Connecticut, 20th century. Eagle finial, brass bezel, painted metal dial, brass side arms, polychrome transfer throat and tablet, molded bracket, housing a brass 8-day spring driven movement.
Length 37 1/2 in. $500
The next clock is a Mahogany Banjo Timepiece, Waterbury Clock Com-

pany, Waterbury, Connecticut, 20th century. Wooden finial, brass bezel, porcelain dial, convex glass, brass side arms, reverse painted throat and tablet, housing a brass 8-day weight driven movement.
Length 41 in. $600
Fifth in line is an Oak Banjo Clock, Waterbury Clock Company, Waterbury, Connecticut, 20th century. A turned finial, brass bezel, porcelain dial, convex glass, brass side arms, gilt transfer throat and tablet, housing a brass 8-day time-and-strike movement.
Length 43 in. $625
Lastly is a Mahogany Banjo Clock, E. Ingraham Clock Company, Bristol, Connecticut, 20th century. Eagle finial, brass bezel, convex glass, metal dial, brass side arms, "Treasure Island" throat and tablet, housing a brass 8-day time-and-strike movement.
Length 38 in. $450
Robert W. Skinner, Inc.

No. 3 Regulator Clock, Welch Spring & Co., Forestville, Connecticut, c. 1880. Walnut veneered case, black and gold tablet with "Regulator" inscribed, painted metal dial, 8-day weight driven movement.
Height 34 1/2 in. $500
American Clock & Watch Museum

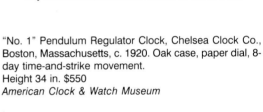

Short Drop Regulator Clock, Seth Thomas Clock Company, Thomaston, Connecticut, c. 1875. Painted zinc dial, black and gold lower glass, 8-day movement, rosewood veneered case. $400

"No. 1" Pendulum Regulator Clock, Chelsea Clock Co., Boston, Massachusetts, c. 1920. Oak case, paper dial, 8-day time-and-strike movement.
Height 34 in. $550
American Clock & Watch Museum

"Empire" style Column and Cornice Shelf Clock, Spencer, Wooster & Co., Salem Bridge, Connecticut, c. 1870. Mahogany case with a single door, the bottom partition is removable, fully turned columns with carved capitals, painted wood dial, reverse painted center glass, replacement lower glass, high quality 8-day brass weight driven movement.
Height 33 1/2 in. $475
Lindy Larson

Regulator Wall Clock, c. 1880s. Veneered case, 8-day movement. $800

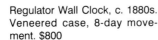

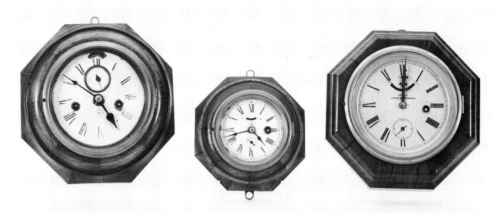

Three gallery clocks, from left to right:
Octagonal Gallery Clock, Seth Thomas Clock Co., Thomaston, Connecticut, c. 1875. Laminated and turned walnut case, painted zinc dial with maker's trademark, paper labels, and an 8-day time-and-strike balance wheel movement.
Height 10 in. $300
Miniature Octagonal Gallery Clock, Ansonia Clock Co., New York, c. 1880. Oak case, zinc dial with fast-slow indicator and seconds hand, signed on dial and paper label, 30-hour balance wheel movement.
Height 6 1/2 in. $250
Octagonal Gallery Clock, S. Emerson Root, Bristol, Connecticut, c. 1870. Rosewood veneered case, zinc dial with applied paper with seconds indicator, 30-hour balance wheel movement.
Height 9 in. $195
Richard A. Bourne Co., Inc.

Scarce Gallery Clock, Seth Thomas Clock Co., Thomaston, Connecticut, c. 1895. "Lobby" Model of carved oak with turned finials, 18 inch diameter painted zinc dial inscribed by the maker with seconds indicator, 15-day double wind time spring movement with pendulum.
Height 37 in. $1100
Richard A. Bourne Co., Inc.

Two lovely Wall Mirror clocks, from left to right:
Rare Inverted Ogee Veneered Mirror Clock, George Hills, Plainville, Connecticut, c. 1842. Fine rosewood veneered case with a mirror in the upper door and one in the lower section. The 30-hour movement is very unusual and features a cast iron back plate, a lever escapement that operates off of two escape wheels, and a single spring that operates both the time-and-strike mechanisms.
Height 25 3/4 in. $450
Rare Inverted Ogee Veneered Mirror Clock, movement by Seth Thomas Clock Co., Thomaston, Connecticut, c. 1850. Beautifully veneered case with a mirror, an upper door with painted glass tablet, 30-hour time-and-strike balance wheel movement.
Height 36 1/4 in. $500
Richard A. Bourne Co., Inc.

Rare New Hampshire Mirror Clock, unsigned, but attributable to Benjamin Morrell, Boscawen, New Hampshire, c. 1820. Typical New Hampshire mirror clock with gilded and painted columns, brass rosettes in the corners, mirror in the lower section, painted tablet in the upper section opening to a fine original dial. Weight driven 8-day brass time-only movement with Benjamin Morrell's "Wheelbarrow" movement, retaining original weight.
Height 30 in. $4500
Richard A. Bourne Co., Inc.

Rare Wall Clock, Joseph Ives, Bristol, Connecticut, c. 1825. Very impressive clock with molded and gilded gesso frame, mirror tablet, painted upper tablet with landscape and classical buildings, 8-day time-and-strike movement with iron plates and rolling pinions, original iron weights, interior of case beautifully sponge painted.
Height 49 in. $1800
Richard A. Bourne Co., Inc.

Rare Wall Mirror Clock, Joseph Ives, Bristol, Connecticut, c. 1818. Large mirror clock with a door containing two tablets and a mirror flanked by reeded columns under bird's-eye maple panels. The door opens to a painted iron dial and exceptional smoke decorated case interior; on top, a scroll with three brass finials. 8-day time-and-strike movement with iron plates, brass gears, roller pinions and original weights.
Height 55 1/2 in. $3000
Richard A. Bourne Co., Inc.

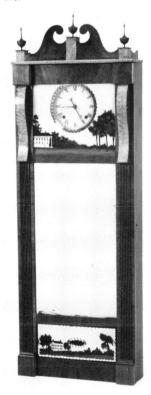

Very Rare Mirror Clock, Joseph Ives, Bristol, Connecticut, c. 1818. Appealing mahogany case with birds-eye panels and tiger maple molding, two painted tablets, a mirror, scrolled crest and three brass finials. Case interior retains old blue paint, painted iron dial, 8-day time-and-strike movement with long pendulum, iron plates, brass gears and roller pinions; retains period weights and pendulum bob.
Height, incl. finial. 56 in. $2750
Richard A. Bourne Co., Inc.

Very Rare Wall Clock, Joseph Ives, Bristol, Connecticut, c. 1820. Constructed of mahogany with scrolled top, a painted tablet and mirror, blue painted case interior, Ives Patent movement with iron back plate, pierced brass front plate, brass gears with roller pinions, and a long pendulum. Exceptional painted iron dial with gold decorations, period inscription on the back, "Bristol, Con't," 3 period brass finials, 2 iron weights.
Height 55 1/2 in. $1300
Richard A. Bourne Co., Inc.

"Star Drop" Victorian Wall Regulator Clock, William L. Gilbert Clock Co.,
Winsted, Connecticut, c. 1875. Walnut case with carved leaves applied to
the unusual ten-sided top, stenciled glass of geometric designs and an owl,
large brass pendulum, paper dial.
Height 32 1/2 in. $850
Lindy Larson

No. 1 Regulator Clock, Welch, Spring & Co., Forestville, Connecticut, c.
1875. Victorian-style rosewood veneered case, full turned columns and
finials, painted metal dial, brass pendulum, 8-day double weight driven
movement.
Height 72 in. $700
American Clock & Watch Museum

Wall Regulator Clock, Boston Clock Co., Boston, Massachusetts, c. 1890.
Solid cherry case, signed painted dial, fancy damascened brass pendu-
lum, high quality signed 8-day weight driven movement.
Height 27 in. $3500
Lindy Larson

Regulator Clock, No. 17, Seth Thomas Clock Co., Thomaston, Connecticut,
c. 1885. Walnut case (these were also made in oak and cherry), "Double
Time" attachment showing local and standard (railway) time, 8-day brass
time-and-strike movement.
Height 76 in. $1000
American Clock & Watch Museum

Regulator Timepiece, E. Howard Watch & Clock Co., Boston, Massachusetts, c. 1884. Oak case, tablet with "Electric Watch Clock," 8-day brass time-and-strike movement. Special ordered in 1884 for the Second National Bank of Cumberland, Maryland, with the addition of a watchman's device.
Height 76 in. $2350
American Clock & Watch Museum

Banjo Timepiece, unsigned, New England, c. 1830-1840. Mahogany case with gold leaf front, applied gold leaf roping and finial, gold leaf and reverse painted glasses, painted iron dial, 8-day weight driven movement.
Height 33 in. $2700
Lindy Larson

Regulator Clock, Eli Terry, Plymouth, Connecticut, c. 1850. Mahogany case, simple style paper dial, 8-day brass time-and-strike movement. Reputed to be the last clock made by Eli Terry after his retirement; it stayed in the hands of his family until donated to the museum.
Height 47 1/2 in. (unable to assign value)
American Clock & Watch Museum

Banjo Clock, unsigned, New England, c. 1870. Banjo style case. eglomise panels, the bottom showing a naval engagement, brass sidearms, paper dial, 8-day time movement.
Height 32 1/2 in. $1600
Winter Associates

Two wall clocks, from left to right:
Rippled Wall Clock, Chauncey Jerome, New Haven, Connecticut, c. 1870. Rosewood veneered case with applied ripple moldings, paper on zinc dial, paper label, unusual 8-day movement with the spring encased in a barrel; movement inscribed by Jerome.
Height 22 in. $300
Rippled School Clock, Chauncey Jerome, New Haven, Connecticut, c. 1860. Fine rosewood veneered case with rippled moldings, painted zinc dial, blue maker's label in bottom of case, entire back board of case is hinged. Rare and unusual 8-day time fusee movement.
Height 21 3/4 in. $750
Richard A. Bourne Co., Inc.

This is a nice example of an Ash Schoolhouse Timepiece, Ansonia Clock Co., Ansonia, Connecticut, c. 1880. It has an octagonal top with brass bezel, long drop, clear glass tablet, housing a brass 8-day movement.
Length 32 in. $500
Robert W. Skinner, Inc.

Two school clocks, from left to right:
Octagon School Clock, Jerome & Company, New Haven, Connecticut, c. 1860. Rosewood veneered case with carved side brackets, lower door with an excellent, original silver and black glass tablet, paper label, painted zinc dial inscribed "Canterbury," 8-day time-and-strike movement. Clock probably made for export.
Height 24 in. $600
School Clock, Seth Thomas Clock Co., Thomaston, Connecticut, c. 1875. Rosewood veneered case with painted gold striping, lower door with original black and gold glass, painted zinc dial, paper label, 8-day time movement inscribed by the maker.
Height 21 1/2 in. $500
Richard A. Bourne Co., Inc.

A classic is this Oak Schoolhouse Timepiece, Seth Thomas Clock Company, Thomaston, Connecticut, c. 1880. It has the octagonal top, wooden bezel, painted metal dial, extra long drop, clear glass tablet, housing a brass 8-day weight driven movement.
Length 53 1/2 in. $750
Robert W. Skinner, Inc.

Three schoolhouse regulator clocks, from left to right:
Schoolhouse Short Drop Regulator, Sessions Clock Co., Bristol, Connecticut, c. 1900. "Century" model, pressed oak case with fine label, 8-day time, strike and calendar movement.
Height 26 1/2 in. $500
Schoolhouse Short Drop Regulator, Seth Thomas Clock Co., Thomaston, Connecticut, c. 1875. Rosewood case with 8-day time-and-strike movement.
Height 22 in. $600
Long Drop Schoolhouse Regulator, Seth Thomas Clock co., Thomaston, Connecticut, c. 1875. "Globe" model, rosewood veneered case, fine original labels, 8-day time-only movement.
Height 31 1/2 in. $675
Richard A. Bourne Co., Inc.

Round Drop Calendar Clock, Ansonia Brass & Copper Co., Ansonia, Connecticut, c. 1875. Rosewood veneered case, turned wood bezel that opens to a perfect dial inscribed by the maker and "Terry's Patent. " The case features choice veneers, black and gold tablet and turned finials. 8-day time-and-strike movement with attached calendar mechanism. Date ring on outside of the dial and the month is indicated by a small dial that revolves behind the time dial.
Height 26 in. $1350
Richard A. Bourne Co., Inc.

Round Drop Calendar Clock, Ansonia Brass & Clock Co., Ansonia, Connecticut, c. 1875. In a rosewood veneered case with a turned wood bezel that opens to a perfect dial inscribed by the maker and "Terry's Patent. " The case features choice veneers, a fine original black and gold tablet, turned finials, fine paper label, date ring on outside of dial and the months indicated by a small dial that revolves behind the time dial. 8-day time-and-strike movement with attached calendar mechanism.
Height 26 in. $1600
Richard A. Bourne Co., Inc.

Scarce Calendar Clock, The Terry Clock Co., Waterbury, Connecticut, c. 1860. In an unusual case with figured mahogany veneers, painted zinc dial with red numerals to indicate the date, painted gold and black tablet, paper label, 8-day time movement marked by the maker.
Height 22 1/4 in. $800
Richard A. Bourne Co., Inc.

The left hand clock is a Figure Eight Calendar Timepiece, E. Ingraham & Co., Bristol, Connecticut, c. 1880. It has a round wooden bezel, paper dial, gold and black tablet, housing an 8-day movement with a simple calendar mechanism.
Length 24, dial dia. 11 1/4 in. $450
The right hand clock is a Pressed Oak Schoolhouse Calendar Timepiece, Sessions Clock Co., Forestville, Connecticut, c. 1880. Note the octagonal top, brass bezel, paper dial, short drops with "Regulator" tablet, housing a brass 8-day movement with a simple calendar mechanism.
Length 27 in. $450
Robert W. Skinner, Inc.

Three scarce Calendar clocks, from left to right:
Scarce Round Drop Calendar Clock, Ansonia Brass & Clock Co., Ansonia, Connecticut, c. 1870. Featuring William Terry's Patent calendar mechanism with an opening in the dial through which the month can be seen, while the date is indicated by a pointer. Rosewood veneered case, exceptional painted glass tablet, paper label, 8-day time-and-strike movement with pendulum.
Height 24 1/2 in. $1500
Very Rare Calendar Clock, Burwell & Carter, Bristol, Connecticut, c. 1870. With B. B. Lewis Perpetual Calendar mechanism with separate day, date and month indicators, painted zinc dial, 8-day time movement with period pendulum, and lower door with a fine black and gold tablet.
Height 24 1/2 in. $2000
Rare Calendar Clock, Ansonia Brass & Clock Co., c. 1865. "Novelty Calendar" with a roll of paper indicating the days of the month through a window in the lower door. Rosewood veneered case, painted zinc dial, 8-day time movement with pendulum, exceptional black and gold tablet.
Height 24 1/2 in. $950
Richard A. Bourne Co. Inc.

Interior view of Stennes banjo clock.
Lindy Larson

Banjo Style Wall Clock, Elmer Stennes, Weymouth, Massachusetts, c. 1971. Mahogany crossbanded and inlaid case with reverse painted glasses, signed dial, weight, movement and case. Very high quality, limited production and highly collectible timepiece.
Height 34 in. $2200
Lindy Larson

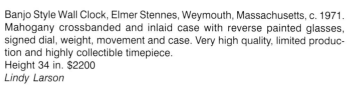

Another interior view of Stennes banjo clock.
Lindy Larson

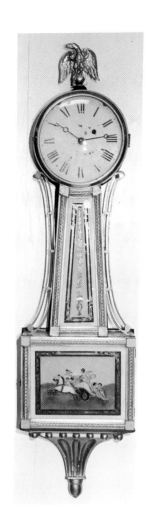

Banjo Clock, Aaron Willard Jr., Boston, Massachusetts, c. 1845. Lovely gilt decorated banjo case, brass sidearms, eagle finial, lower panel of horse and chariot, red and gold decorated upper panel, 30-hour brass time movement. Height 33 in. $1475
Winter Associates

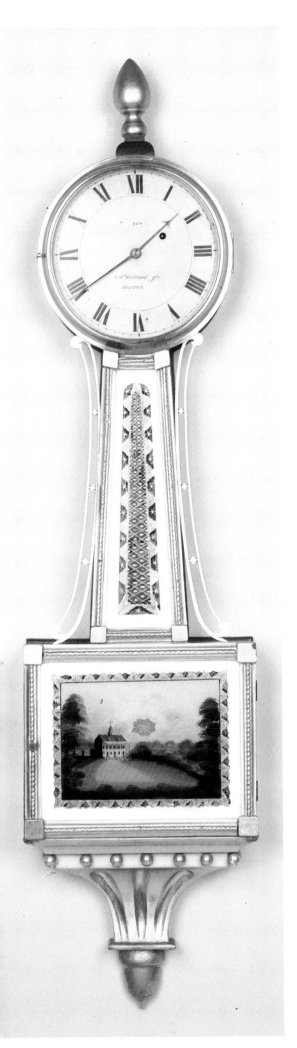

Federal Gilt Gesso and Mahogany Banjo Timepiece, Aaron Willard, Jr., Boston, Massachusetts, c. 1820. The circular brass molded bezel enclosing a painted iron dial inscribed "2136 Aaron Willard Jr. BOSTON" and 8-day weight driven brass movement above the tapering throat and rectangular pendulum box with eglomise tablets, the lower depicting Mount Vernon, both framed by spiral gilt moldings flanked by pierced brass side arms, the whole resting on a giltwood flared bracket ending in acorn pendant.
Height 40 in. $60,000
Robert W. Skinner, Inc.

"Model #2" Banjo Clock, Howard & Davis, Boston, Massachusetts, c. 1850. Case grained to simulate rosewood, 10 inch diameter dial, 30-hour time movement. Banjo style clocks were made in five different sizes.
Height 44 in. $1000
American Clock & Watch Museum

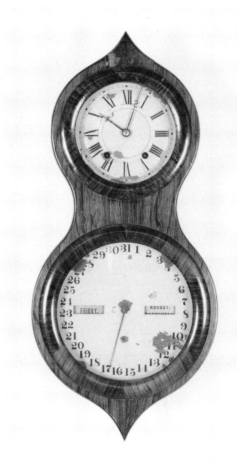

Immaculate Original Condition Office Wall Calendar Clock, Seth Thomas Clock Co., Thomaston, Connecticut, c. 1875. Rosewood veneered case, 8-day time-only movement, calendar on the bottom.
Height 32 1/2 in. $800
Richard A. Bourne Co., Inc.

A very rare "Peanut" Model Calendar Wall Clock, Seth Thomas Clock Co., Thomaston, Connecticut, c. 1875. Rosewood veneered case, 8-day double wind, time-only movement, "S. T. " hands.
Height 23 1/4 in. $750
Richard A. Bourne Co., Inc.

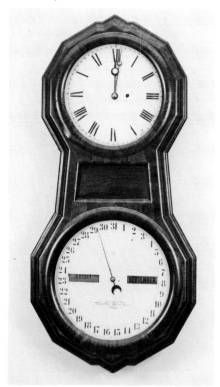

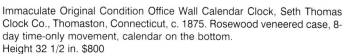

Office Calendar Clock, Seth Thomas Clock Co., Thomaston, Connecticut, c. 1875. Office Calendar #1 with an exceptional rosewood veneered case with a hinged front. Two painted zinc dials, both in mint condition, the lower with the Seth Thomas trademark and "Patented Feb. 15, 1876. " On the inside of the door is the original instruction label. 8-day weight driven time movement with a gridiron pendulum and a period cast iron weight. Calendar mechanism indicates date, day and month.
Height 40 3/4 in. $1550
Richard A. Bourne Co., Inc.

Walnut Double Dial Calendar Wall Timepiece, Seth Thomas Clock Co., Thomaston, Connecticut, c. late 19th century. A variant of the Office Calendar No. 6, painted black.
Length 32 in. $1500
Robert W. Skinner, Inc.

A Rosewood Veneer Double Dial Calendar Timepiece, Seth Thomas Clock Company, Thomaston, Connecticut, c. mid-19th century. This is the "Office Calendar No. 1," with painted metal dials, housing an 8-day weight driven movement.
Length 40 in. $1450
Robert W. Skinner, Inc.

Double Dial Calendar Clock, Waterbury Clock Co., Waterbury, Connecticut, c. 1889. Model #33, oak case with pressed and reeded decoration, turned finials, two painted zinc dials both inscribed by the maker, and the calendar mechanism. Paper labels, 8-day time-and-strike movement with a large brass bob.
Height 39 1/2 in. $1400
Richard A. Bourne Co., Inc.

No. 6 Hanging Library Calendar Clock, Ithaca Calendar Clock Co., Ithaca, New York, c. 1881. Excellent walnut case with a bracket, nice moldings, carved crest, signed label and calendar dial, 8-day time-and-strike movement, and a calendar dial and mechanism in the bottom.
Height 33 in. $1350
Richard A. Bourne Co., Inc.

Oak Double Dial Calendar Wall Clock, Waterbury, Connecticut, c. 1898. Spool gallery, reeded door, painted metal dials, applied bracket, housing an 8-day time-and-strike movement.
Length 35 in. $1400
Robert W. Skinner, Inc.

Marine Gallery Timepiece, Charles Kirk, New Haven, Connecticut, c. 1847. Walnut case with a painted zinc dial.
Dia. 11 in. $700
Lindy Larson

Interior view of the Charles Kirk Marine Gallery timepiece showing the typical Kirk horizontal movement. There is a second escape wheel behind the one visible in the photo, thus the term "double escape wheel" is used when referring to these models; they usually run somewhat longer than 30 hours.
Lindy Larson

Marine Gallery Clock, Marine Clock Manufacturing Co., New Haven, Connecticut, c. 1847. Wood case with a brass front stamped with the maker's name, painted zinc dial, label, on the back of the case reads: "Marine timepieces manufactured by the Marine Clock Manufacturing Co., New Haven, Conn. Patented by Chas. Kirk, April, 1847 U. S. A. "
Dia. 9 1/2 in. $1450
Lindy Larson

Interior view of the Marine Gallery Clock. Some unusual features of this 8-day timepiece are the double escape wheels and the remote chain driven fusee barrel.
Lindy Larson

New Hampshire Mirror Clock, unsigned, c. 1825. Figured mahogany reverse ogee case with flame mahogany corners, reverse painted upper glass, painted iron dial lead weight, 8-day timepiece.
Height 33 in. $4500
Lindy Larson

Octagon School Clock, Chauncey Jerome, New Haven, Connecticut, c. 1855. Hardwood case is painted black, with black and gold decorated upper and lower glasses, octagon dial is of painted zinc, 80 day timepiece.
Height 18 in. $500
Lindy Larson

Interior view of the Chauncey Jerome Octagon School Clock.
Lindy Larson

Perpetual Calendar Clock, Welch, Spring & Co., Forestville, Connecticut, c. 1875. Rosewood veneered case, gold decorated tablet, paper on metal dial, 8-day time-and-strike weight and spring driven movement.
Height 31 in. $1000
American Clock & Watch Museum

Interior view of the New Hampshire Mirror Clock.
Lindy Larson

Double Dial Calendar Wall Clock, Seth Thomas, Plymouth Hollow, Connecticut, c. 1860. Office calendar No. 3 commonly called the "Peanut" model; rosewood case, painted zinc dials, 5 inch time dial, 7 inch calendar dial. 8-day double wind timepiece.
Height 23 1/2 in. $3400
Lindy Larson

Interior view of the Double Dial Calendar Wall Clock, showing the solid plate double wind time movement and the Seth Thomas calendar movement.
Lindy Larson

Two rare calendar clocks, from left to right:
Rare Calendar Clock, Gilbert Manufacturing Co., Winsted, Connecticut, c. 1861. With Galusha Maranville Patent Calendar Mechanism. Mahogany veneered case, painted zinc dial, painted glass tablet, paper label, 8-day time-and-strike spring movement; calendar mechanism gives month, date and day of the week. The dial is made of three parts, so that the rear dials can be turned manually to line up the correct day of the week to the correct date of the month.
Height 24 1/2 in. $900
Scarce Gallery Clock, Seth Thomas Clock Co., Thomaston, Connecticut, c. 1890. Mahogany case, painted zinc dial with seconds indicator, 30-day double wind time movement inscribed by the maker with dead beat escapement.
Height 19 in. $600
Richard A. Bourne Co., Inc.

A lovely example of a Rosewood Veneer Double Dial Calendar Timepiece, E. N. Welch Manufacturing Company, Forestville, Connecticut, c. 1865. This clock is similar to the Welch Spring "Regulator Calendar No. 1," housing a brass 8-day double weight drive movement with a B. B. Lewis calendar mechanism.
Length 48 in., dial dia. 13 1/2 in. $2800
Robert W. Skinner, Inc.

Two outstanding calendar wall clocks, from left to right:
Calendar Wall Clock, Ingraham Clock Co., Bristol, Connecticut, c. 1870-1880. Figure "8" rosewood case, fine labels, 8-day time-and-strike movement, B. B. Lewis calendar.
Height 30 in. $675
Calendar Wall Clock, L. F. & W. Carter, c. 1860s. Rosewood veneered case, fine labels, 8-day time-only movement, B. B. Lewis calendar.
Height 30 3/4 in. $700
Richard A. Bourne Co., Inc.

The first clock on the left is a Rosewood Double Dial Clock, Seth Thomas, Thomaston, Connecticut, c. 1880. Painted metal dials, housing a brass 8-day time-and-strike movement.
Length 27 in. $1100

The second clock from the left is a Rosewood Grain-painted Calendar Timepiece, Lovell Manufacturing Company, Erie, Pennsylvania, c. 1880. Round drop, paper dial, housing an 8-day time and simple calendar movement.
Length 24 in., dial dia. 11 3/8 in. $550

The third clock is a Figure Eight Calendar Timepiece, E. Ingraham Company, Bristol, Connecticut, c. 1880 "Mosaic" Model, paper dial, housing an 8-day spring driven movement and a B. B. Lewis calendar mechanism. One rosette is missing.
Length 29 in. $795

The fourth clock is a Gilt Figure Eight Calendar Clock, E. Ingraham & Company, Bristol, Connecticut, c. 1880. "Ionic" Model, paper dial, housing a brass 8-day time-and-strike movement with a B. B. Lewis calendar mechanism.
Length 30 in. $1400
Robert W. Skinner, Inc.

Rare Lyre Clock, unsigned, Massachusetts, c. 1825. A true lyre clock without the bottom door, featuring a beautifully carved throat with two mahogany panels and a fine carved finial, all on a molded base with a bracket. Painted iron dial with 8-day brass time-only movement.
Height 40 1/2 in. $5000
Richard A. Bourne Co., Inc.

Lyre Clock, maker unknown, c. first quarter nineteenth century. Exceptional eagle, cornucopia door glass and lyre throat glass; a true box lyre of the period with mahogany sides and faces and presentation piece. 8-day time-only movement.
Height 40 1/2 in. $6500
Richard A. Bourne Co., Inc.

Calendar Clock, Seth Thomas Clock Co., Thomaston, Connecticut, c. 1876. Large mahogany veneered case, painted metal dials, 8-day time-and-strike movement.
Height 38 in. $800
American Clock & Watch Museum

Calendar Clock, Welch, Spring & Co., Forestville, Connecticut, c. 1885. Ornate oak case, full turned columns and finials, elaborate headpiece, painted metal dials, 8-day time-and-strike movement.
Height 39 in. $950
American Clock & Watch Museum

Wall Clock, Williams & Hatch, North Attleboro, Massachusetts, c. 1850. Cherry case stained to appear like rosewood, black and gold tablet, painted metal dial, 8-day weight driven movement.
Height 33 in. $750
American Clock & Watch Museum

"Kosmic" #70 Model Clock, E. Howard & Co., Boston, Massachusetts, c. 1880. Rosewood veneered case, red, gold and black tablet, painted metal dial, 8-day time-only movement. This clock was made to provide 24 hour time by changing Roman Numerals I - XII to Arabic numbers 13 through 24 beginning at noon. This idea did not prove to be popular with the public.
Height 30 in. $650
American Clock & Watch Museum

Howard #8 Figure "8" Wall Clock, Howard & Co., Boston, Massachusetts. Superb black walnut case, with 8-day weight driven brass time-only movement.
Height 44 1/2 in. $6000
Richard A. Bourne Co., Inc.

Two outstanding wall clocks, from left to right:
Rare Octagon Wall Clock, Atkins Whiting & Co., Bristol, Connecticut, c. 1850-1854. Rosewood veneered case with applied rippled moldings, original tablet in untouched condition, very rare feature of Joseph Ives Patent 30-day lever spring movement.
Height 25 1/4 in. $900
Rare Calendar Clock with Galusha Maranville Patent Calendar Mechanism, Gilbert Manufacturing Co., Winsted, Connecticut, c. 1861. Mahogany veneered case, paper label, painted glass tablet, 8-day brass time-and-strike movement, dial reads, "Patented by Galusha Maranville, March 5, 1861. " Calendar gives month, date and day of the week; dial is made in two parts so that the rear dial can be turned manually each month to line up the correct day of the week to the correct date of the month.
Height 24 1/2 in. $795
Richard A. Bourne Co., Inc.

Four clocks from the 20th century, from left to right:
Schoolhouse Clock, E. Ingraham & Co., Bristol, Connecticut, c. 1900. Fine pressed oak case, 8-day time, strike and calendar movement.
Height 18 3/4 in. $350
Schoolhouse Clock, Sessions Clock Co., Bristol, Connecticut, c. 1915-1920. Oak case, 8-day time-and-strike and calendar movement, retaining part of original label.
Height 19 1/2 in. $375

Miniature Kitchen Wall Clock, New Haven Clock Co., New Haven, Connecticut, c. 1930s. Original white painted case, 8-day time-only movement, excellent condition.
Height 11 3/4 in. $300
Schoolhouse Clock, Seth Thomas Clock Co., Thomaston, Connecticut, c. 1920s. Oak case, 8-day time-only movement, original condition.
Height 20 in. $325
Richard A. Bourne Co., Inc.

Three outstanding wall clocks, from left to right:
Long Drop Regulator, Ansonia Clock Co., Ansonia, Connecticut, c. 1900. "Regulator A" with a walnut case with ebonized trim, paper on zinc dial, tablet with silver lettering and an 8-day time-and-strike movement.
Height 32 in. $600
School House Clock, Ansonia Clock Co., Ansonia, Connecticut, c. 1900. "Drop Octagon B. " Molded oak case, paper on zinc dial, black and gold tablet, maker's label, 8-day time-and-strike movement with half-hour strike.
Height 26 in. $500
Long Drop Regulator, New Haven Clock Co., New Haven, Connecticut, c. 1900. "Bank Regulator" with a fancy pressed oak case, lower glass with gold lettering, maker's label, 8-day time movement.
Height 36 in. $600
Richard A. Bourne Co., Inc.

Three choice wall clocks, from left to right:
Wall Clock, E. Ingraham & Co., Bristol, Connecticut, c. 1880. "Ionic" Model with a grained poplar case, original black and gold table, paper on zinc dial, paper label and an 8-day time-and-strike movement.
Height 21 1/2 in. $700
Very Rare Calendar Clock, Boardman & Hubbell, Forestville, Connecticut, c. 1867. Said to be one of two known examples. Walnut case with ebonized trim featuring an unusual 8-day lever movement marked "L. Hubbell," and Alfonzo Boardman's Patent calendar mechanism employing one roller for month and date, with day of the week on another roller. A 30-hour spring operates the calendar mechanism for one year; month and date strip is made of sized linen with a black and gold tablet in front of them. Painted zinc dial.
Height 27 in. $2000
Regulator Clock, Atkins Clock Co., Bristol, Connecticut, c. 1870. Rosewood veneered case with a black and gold tablet, paper label, painted zinc dial and an 8-day time-and-strike movement.
Height 24 1/4 in. $750
Richard A. Bourne Co., Inc.

Two wall clocks, from left to right:
Calendar Clock, Ithaca Calendar Clock Co., Ithaca, New York, pat. Aug. 28, 1866. "No. 4 Hanging Office" Model, rosewood veneered case, paper label, two paper on zinc dials, the lower with date indicator, and day and month apertures, scarce 30-day nickel plated double spring time movement, manufactured by E. N. Welch for the Ithaca Calendar Clock Co.
Height 28 3/4 in. $1100
Wall Clock, "Office #1," Seth Thomas, Thomaston, Connecticut, c. 1865. Rosewood veneered case, black and gold stenciled glass, painted zinc dial with company trademark, paper label, 8-day time movement stamped by Seth Thomas Clock Co.
Height 25 in. $600
Richard A. Bourne Co., Inc.

Two wall clocks, from left to right:
Calendar Clock, Seth Thomas Clock Co., Thomaston, Connecticut, c. 1875. "Office Calendar #71/2" walnut veneered case with brass rims around the glasses and applied rosettes, lower dial with day and month apertures and a date indicator, upper dial marked "Patented Dec. 28 1875," and the lower dial marked "Patented Feb. 15, 1876. " 8-day time-and-strike movement.
Height 26 1/2 in. $950
Ionic Wall Clock, E. Ingraham Clock Co., Bristol, Connecticut, c. 1870. Rosewood veneered case with the unusual and scarce feature of alternating light and dark wood on the bezels and rosettes, lower door with black and gold tablet, paper label, paper on zinc dial, 8-day time-and-strike movement.
Height 22 in. $675
Richard A. Bourne Co., Inc.

Two styles of wall clock, from left to right:
Double Dial Calendar Clock, E. N. Welch Manufacturing Co., Forestville, Connecticut, c. 1870. Round Head Calendar #4, rosewood veneered case, upper dial indicates the time and day of the week, lower dial indicates the date and the month, (Note: calendars are written in Spanish) the calendar mechanism mounted behind the lower door is a B. B. Lewis "Y" mechanism. 8-day time-and-strike movement marked by the maker with patent date.
Height 32 1/2 in. $800
Wall Clock, Waterbury Clock Co., Waterbury, Connecticut, c. 1870. Large model figure 8 wall clock with a rosewood veneered case, superb red and gold tablet in the lower section, painted zinc dial, bezels are laminated and turned oak, 8-day time movement inscribed by the maker.
Height 30 in. $500
Richard A. Bourne Co., Inc.

Two different wall clocks, from left to right:
Calendar Clock, Ingraham Clock Co., Bristol, Connecticut, c. 1870. Rosewood veneered case, upper dial with day indicator, lower dial with date and month indicator, painted zinc dial, 8-day movement.
Height 30 in. $950
Ionic Wall Clock, Ingraham Clock Co., Bristol, Connecticut, c. 1868. Rosewood veneered case with exceptional and original gold floral tablet on a blue background, fine paper label, painted zinc dial and an 8-day time movement marked by the maker.
Height 22 in. $875
Richard A. Bourne Co., Inc.

From left to right, the following clocks are:

Short Drop Regulator, Henry Sperry & Co., New York c. 1850. Mahogany veneered case with cross banding, carved brackets, fine painted tablet, paper label, enameled zinc dial, 8-day time movement with pendulum.
Height 24 1/2 in. $500

The middle clock is a lovely Regulator Clock, Silas B. Terry, Terryville, Connecticut, c. 1840. A fine rosewood veneered case with black and gold tablet, enameled zinc dial with seconds indicator, black and gold paper label on weight pan, 8-day time movement with dead beat escapement, maintaining power, gilded pendulum rod, and lead weight.
Height 37 1/2 in. $750

On the right is a Round Drop Calendar Clock, E. Ingraham & Co., Bristol, Connecticut, c. 1870. Poplar case with grain-painted decoration, paper-on-zinc dial, 8-day time movement with simple calendar mechanism and pendulum.
Height 24 in. $400
Richard A. Bourne Co., Inc.

A grouping of three wall clocks, from left to right:

Figure "8" Wall Clock, E. Ingraham & Co., Bristol, Connecticut, c. 1870. Rosewood veneered case, 8-day time-and-strike movement.
Height 22 in. $600

Early Short Drop Regulator, Atkins Clock Co., Bristol, Connecticut, c. 1850. Outstanding clock with mahogany veneered case with applied carving, fine original label, 8-day time-only movement.
Height 25 1/2 in. $550

Short Drop Regulator, c. 1880. Case is English with mahogany veneer, fine original label, 8-day time-and-strike movement by Waterbury Clock Co., Waterbury, Connecticut.
Height 28 1/2 in. $750
Richard A. Bourne Co., Inc.

Two wall clocks, from left to right:
Wall Regulator, possibly by the Waterbury Clock Co., Waterbury, Connecticut, c. 1870. Rosewood veneered case with an excellent black and gold glass, painted zinc dial, 8-day time weight driven cast iron weight. Height 35 in. $850
Victorian Wall Clock, Seth Thomas Clock Co., Thomaston, Connecticut, c. 1880. "Queen Anne" Model, walnut case with turned columns and finials, painted zinc dial, 8-day time-and-strike movement inscribed by the maker, strikes on a bell and retains a nickel plated pendulum bob.
Height 36 3/4 in. $957
Richard A. Bourne Co., Inc.

Two fine wall clocks, from left to right:
Rare Wall Clock, unsigned, probably by George Hatch, North Attleboro, Massachusetts, c. 1850. Poplar case with exceptional grained decoration, two black and gold glasses, painted zinc dial, 8-day weight driven movement retaining a period iron weight.
Height 33 in. $1450
Wall Regulator, E. Howard Clock Co., Boston, Massachusetts, c. 1880. No. 70-5 Model with walnut case, black, red and gold glass, painted dial inscribed by the maker, an 8-day time movement inscribed by the maker, and number "70." Retains an old lead weight.
Height 32 1/4 in. $1150
Richard A. Bourne Co., Inc.

From left to right:
Rare Calendar Clock, Calendar mechanism patented by Galusha Maranville, Movement by William L. Gilbert Clock Co., Winsted, Connecticut, c. 1861. Rosewood veneer case with delicate carved side arms, painted zinc dial with apertures for the month and the day that are manually adjusted, while a simple calendar mechanism activates the date indicator. 8-day time weight driven movement with dead beat escapement, period pendulum and iron weight, exceptional black and gold tablet with inscription "Regulator."
Height 34 in. $2150
#3 Regulator, Seth Thomas, Thomaston, Connecticut, c. 1875. Walnut case veneered with burl walnut, painted zinc dial with seconds indicator inscribed in Old English, "Seth Thomas," 8-day time movement with period pendulum and iron weight.
Height 42 1/2 in. $1600
Richard A. Bourne Co., Inc.

Three clocks, from left to right:
#2 Regulator Clock, Seth Thomas, Thomaston, Connecticut, c. 1890. Walnut case with laminated and turned bezel, label, painted zinc dial inscribed by the maker with seconds indicator, 8-day time weight driven movement with pendulum and brass weight.
Height 36 1/2 in. $1200
In the middle, a scarce Banjo Clock, Brewster & Ingraham, Bristol, Connecticut, c. 1850. Variant of the Brewster & Ingraham Gallery Clock constructed of rosewood and mahogany, painted zinc dial, black and gold glasses, 8-day time movement with pendulum. Retains over-pasted label of J. J. Beals, Boston.
Height 31 1/2 in. $800
On the right, a #2 Regulator Clock, William L. Gilbert & Co., Winsted, Connecticut, c. 1855. Early model with rosewood veneered case, exceptional tablet, painted zinc dial with seconds indicator, 8-day time weight driven movement with lead weight and pendulum. Blue paper label.
Height 33 1/2 in. $950
Richard A. Bourne Co., Inc.

Some lovely examples of wall clocks are shown, left to right, with the first timepiece being a Rosewood Veneer Regulator Timepiece, Seth Thomas Clock Company, Thomaston, Connecticut, c. 1845. This is the "No. 1 Extra Bank Model," with painted dial, gilt "Regulator" tablet, housing a brass 8-day weight driven movement.
Length 40 in. $1200
The second clock from the left is an Oak Wall Clock, Waterbury Clock Company, Ansonia, Connecticut, c. 1900. This is the "Atlanta" model with carved crest, silvered metal dial, fluted columns over conforming free standing columns, housing a brass 8-day pull-up weight driven movement.
Length 52 1/2 in. $850
The center clock is a Walnut Double Dial Calendar Timepiece, Ithaca Calendar Clock Company, Ithaca, New York, c. 1866. This is a "No. 2 Regulator," with painted metal dials, housing a brass 8-day double weight driven movement. There has been minor restoration.
Length 43 1/2 in. $1400
The fourth clock is an Oak Double Dial Calendar Timepiece, New Haven Clock Company, New Haven, Connecticut, c. 1880. This is the "Rutland" model, with shaped crest with applied rosettes, baluster shaped bosses and leaves, paper dials, housing a brass 30-day double spring movement and simple calendar mechanism.
Length 48 in. $1250
The last clock is an Oak Regulator Timepiece, Waterbury Clock Company, Waterbury, Connecticut, c. 1880. This is the "Springfield" model, pierced and scrolled crest, paper dial, clear door glass, housing a brass dial 8-day movement.
Length 39 3/4 in. $650
Robert W. Skinner, Inc.

Three wonderful wall clocks, from left to right:
Model #1 Calendar Wall Clock, Seth Thomas Clock Co., Thomaston, Connecticut, c. 1860s. Mahogany case with good label on back of door, 8-day weight driven time-only movement with calendar at bottom.
Height 41 1/4 in. $725
Outstanding Model #18 Long Drop Wall Clock, Seth Thomas Clock Co., Thomaston, Connecticut, c. 1875. Quarter grain oak veneered case, 8-day weight driven time-only movement with second bit.
Height 54 in. $700
Model #13 Regulator Wall Clock With Calendar, Seth Thomas, Thomaston, Connecticut, c. 1875. Pressed oak case, 8-day weight driven time-only movement with second bit.
Height 48 in. $675
Richard A. Bourne Co., Inc.

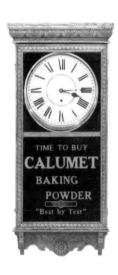

Three turn of the century clocks are from left to right;
A Pressed Oak Regulator Timepiece, Waterbury Clock Company, Waterbury, Connecticut, c. 1880. The crest is shaped, black and gold reverse painted tablets, "Regulator," housing a brass 8-day movement.
Length 37 in. $450
Next is a Pressed Oak Box Timepiece, E. N. Welch Manufacturing Company, Forestville, Connecticut, c. 1900. It has a corniced top, door frame impressed with five-pointed stars, paper dial, black and gold tablet "Time to Buy/Calumet/Baking/Powder/'Best by Test'," housing a brass 8-day movement.

Length 37 in. $400
The clock on the right is a Pressed Oak Box Timepiece, William L. Gilbert Clock Company, Winsted, Connecticut, c. 1900. It has a shaped crest with applied medallion, embossed door frame, gilt and black reverse painted tablets, "Regulator," housing a brass 8-day movement.
Length 37 in. $475
Robert W. Skinner, Inc.

Three assorted wall clocks, from left to right:
Wall Regulator "Observatory" Model, William L. Gilbert Clock Co., Winsted, Connecticut, c. 1880. Solid oak case with two original tablets, paper on tin dial, paper label, 8-day time-and-strike movement.
Height 32 in. $300
Wall Calendar Regulator, Sessions Clock Co., Bristol, Connecticut, c. 1880. Poplar case with two original tablets, paper on tin dial, 8-day time movement with calendar mechanism.
Height 32 in. $350
Wall Clock, E. Ingraham Clock Co., Bristol, Connecticut, c. 1920. Softwood case with paper dial inscribed by the maker and "Forestville, Toronto," 8-day time movement with fancy pendulum.
Height 33 in. $200
Richard A. Bourne Co., Inc.

Three fine wall clocks, from left to right:
#70 Regulator, E. Howard & Co., Boston, Massachusetts, c. 1890. Exceptional quality throughout with a solid oak case, painted zinc dial inscribed by the maker, painted glass tablet, 8-day time movement inscribed "E. Howard & Co., Boston, #70" with original weight and pendulum.
Height 31 3/4 in. $1350
Scarce Wall Clock, Seth Thomas Clock Co., Thomaston, Connecticut, c. 1890. Fine oak case with pediment top and carved bracket, painted zinc dial with seconds indicator, 15-day double wind spring movement with original pendulum.
Height 39 1/2 in. $1100
Wall Regulator, Chelsea Clock Co., Boston, Massachusetts for Daniel Pratt's Son. c. 1900. Solid oak with painted zinc dial inscribed "Daniel Pratt's Son, Boston," 8-day weight driven time movement with period pendulum.
Height 33 in. $850
Richard A. Bourne Co., Inc.

Three wall clocks, from left to right:
Regulator, Seth Thomas Clock Co., Thomaston, Connecticut, c. 1885. Mahogany and mahogany veneered case, painted zinc dial, paper label, 8-day time movement.
Height 33 in. $950
Scarce Inverted Ogee Mirror Clock, movement with Charles Kirk patent, made and sold by George Hills, Plainville, Connecticut, c. 1840. Fine rosewood veneered case with a mirror in the lower section, black and gold decorated glass in the upper door; 30-hour time-and-strike movement with a lever escapement that operates off of two escape wheels, and has a single spring that operates both the time-and-strike mechanisms.
Height 19 1/4 in. $625
Long Drop Regulator, New Haven Clock Co., New Haven, Connecticut, c. 1875. Walnut case, gold decorated glass tablet, painted zinc dial inscribed "New Haven" and an 8-day time spring driven movement.
Height 32 1/2 in. $600
Richard A. Bourne Co., Inc.

Three wall clocks are, from left to right:
Regulator Clock, F. Kroeber & Co., New York, c. 1880. Vienna Regulator #51 with slender case, glazed case sides, porcelain dial with seconds indicator, 8-day time movement stamped by Kroeber with brass weight and pendulum.
Height 44 in. $650
Scarce Oak Wall Clock, E. Ingraham & Co., Bristol, Connecticut, c. 1880. Carved and pressed case, paper on zinc dial, 8-day time movement with pendulum.
Height 37 1/2 in. $550
Regulator Clock, Seth Thomas, Thomaston, Connecticut, c. 1890. Constructed of light mahogany with enameled zinc dial with seconds indicator inscribed by the maker, 8-day time movement with brass weight and pendulum.
Height 36 1/2 in. $975
Richard A. Bourne Co., Inc.

Two wall clocks, from left to right:
Rare Lyre Clock, unsigned, Massachusetts, c. 1830. Mahogany case with a bracket and pendulum opening in the lower door, painted zinc dial. Has the rare feature of a time-and-alarm 8-day weight-driven movement sounding the alarm on a bell on top of the case.
Height 37 1/2 in. $1150
Banjo Clock, unsigned, American, c. 1920. Mahogany case with two fine eglomise panels, brass side arms, brass bezel and a brass eagle finial, porcelain dial, quality 8-day time movement with lead weight and pendulum.
Height 33 in. $1000
Richard A. Bourne Co., Inc.

Two attractive wall clocks, from left to right:
Scarce Lyre Clock, dial inscribed "Levi Hutchins," c. 1825. Mahogany veneered case with a carved throat frame and bird's-eye maple panels, painted iron dial, small 8-day time weight driven movement with plates measuring only 3-1/8 inches x 2-1/4 inches.
Height (Incl. finial) 33 in. $1300
Banjo Clock, unsigned, New England, c. 1830. Mahogany and mahogany veneered case with brass side arms, painted dial, two fine eglomise panels, 8-day time movement retaining a lead weight and pendulum rod.
Height 33 in. $1250
Richard A. Bourne Co., Inc.

Two separate wall clocks, from left to right:
Exceptional New Hampshire Mirror Clock, Benjamin Morrell, Boscawen, New Hampshire, c. 1816-1845. Rare wheelbarrow 8-day time-only weight driven movement, winds at 10 o'clock; retains most of the original label.
Height 30 1/4 in. $6400
Banjo Clock, Tifft, c. 1830-1840. Mahogany case, 8-day time-only movement.
Height 34 in. $900
Richard A. Bourne Co., Inc.

Two wall clocks, from left to right:

Mirror Clock, attributed to I. Dewey, Chelsea, Vermont, c. 1830. Pine case with a gilded door, painted tablet, painted iron dial, 8-day skeletonized movement. This case has a cut-out where the alarm bell was, and the alarm mechanism is now missing.
Height 30 in. $750

Ogee Clock, Chauncey Jerome, New Haven, Connecticut, c. 1860. Standard rosewood veneered ogee case, exceptional floral tablet, painted zinc dial, maker's label on backboard, 30-hour time-and-strike movement.
Height 25 3/4 in. $375
Richard A. Bourne Co., Inc.

Standard Electric Wall Master Clock, Standard Electric Time Co., Inc., c. 1915. Quarter grained oak case, mercury pendulum.
Height 73 in. $1450
Richard A. Bourne Co., Inc.

Standard Electric Wall Master Clock, Standard Electric Time Co., Inc., c. 1915. Quarter grained oak case, nickel pendulum.
Height 74 in. $1000
Richard A. Bourne Co., Inc.

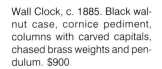

Wall Clock, c. 1885. Black walnut case, cornice pediment, columns with carved capitals, chased brass weights and pendulum. $900

Watchman's Clock, Howard Co., Boston, Massachusetts, c. 1880. 8-day weight driven time-only movement.
Height 54 in. $1100
Richard A. Bourne Co., Inc.

Mahogany Wall Timepiece, labeled G. A. Jones & Co., Cortland Street, New York, c. 1870. "Fritz" case with turned finial over arched top, brass bezel, painted metal dial, free standing baluster shaped columns, square door stenciled in gilt "14," shaped pediment with centering column terminating in a turned pendant, housing an unsigned 8-day brass movement.
Length 48 in. $1050
Robert W. Skinner, Inc.

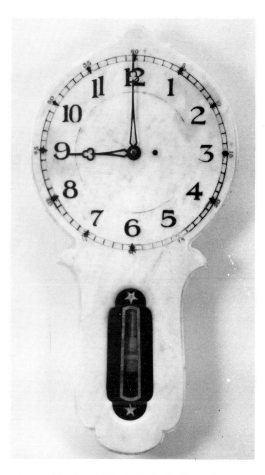

Scarce Marble Dial Wall Clock, E. Howard Clock Co., Boston, Massachusetts, c. 1855. #27 Model, pine case with white enamel, marble front with a black and gold tablet. Fine quality 8-day time movement marked "E. Howard, Boston, #27," retains the original weight and gilded pendulum rod.
Height 35 1/2 in. $1850
Richard A. Bourne Co., Inc.

Rare Wall Acorn Clock, Chauncey Boardman, Bristol, Connecticut, c. 1850. A fine mahogany veneered case with grain-painted sides, two doors; the upper opening to a painted dial and the lower with a painted tablet. Rare 30-hour time-and-strike fusee movement. Maker's signature on movement.
Height 28 1/4 in. $1500
Richard A. Bourne Co., Inc.

Two lovely, rare clocks, from left to right:
Rare Wall Acorn Clock, J. C. Brown/Forestville Manufacturing Co., Forestville, Connecticut, c. 1850. Unusual laminated case with figured mahogany, painted glass tablet, enameled zinc dial, paper label, 8-day time movement stamped "Forestville Mfg. Co., Bristol, U. S. A. " mounted on a seat board with pendulum.
Height 28 1/2 in. $950
Very Rare Calendar Clock, E. Ingraham Clock Co., Bristol, Connecticut, c. 1868. Featuring Josiah K. Seem's calendar dial consisting of three small discs attached to the back of the dial indicating date, day and month through apertures in the dial. Rosewood veneered case with laminated bezels, fine painted tablet depicting a hot air balloon and flags, paper label, enameled zinc dial inscribed "Seem's dial, Patented Jan. 7, 1868. " 8-day time-and-strike movement with pendulum.
Height 22 in. $1200
Richard A. Bourne Co., Inc.

On the left is a Rosewood Veneer "Peanut" Calendar Timepiece, Seth Thomas Clock Company, Thomaston, Connecticut, c. 1866. It has painted metal dials, original "S" and "T" hands, 8-day brass double spring wound time movement with month, day and date calendar mechanism.
Length 23 1/2 in. $3200
The right hand clock is a Walnut and Burl Veneer Gingerbread Wall Clock, Wm. L. Gilbert Clock Company, Winsted, Connecticut, c. 1880. Note the arched crest with shield embossed center plinth with turned finial, painted dial, silver gilt decorated tablet, housing a brass 8-day time-and-strike movement.
Length 28 1/2 in. $500
Robert W. Skinner, Inc.

Three fine wall clocks, from left to right:
Rare Banjo Clock, probably by Eli Terry & Sons, or Henry Terry, Plymouth, Connecticut, c. 1830. Unusual square case with mahogany veneer, two fine eglomise tablets, painted wood dial, 30-hour time-and-strike movement with iron weights and an old pendulum.
Height 34 3/4 in. $1200
Rare Banjo Clock, Brewster & Ingraham, Bristol, Connecticut, c. 1860. Mahogany and rosewood case with two black and gold eglomise tablets, painted zinc dial, 8-day time movement. This is considered the earliest form

of a spring-driven Banjo clock.
Height 31 1/4 in. $550
Scarce Wall Clock, unsigned, dial inscribed "HN Lockwood, Boston," c. 1870. Walnut case with carved decorations, black and gold eglomise tablet, painted zinc dial, 8-day time Banjo-type movement, an old weight and pendulum.
Height 35 1/2 in. $600
Richard A. Bourne Co., Inc.

Two unusual and rare wall clocks, from left to right:
Very Rare Iron Front Wall Clock, Samuel Botsford, Hamden, Connecticut, c. 1853. A very unusual and possibly unique iron front with colorful decorations and an octagon wood case with exceptional wood grained paint, painted zinc dial, and a very rare 8-day movement with Botsford Patent marine escapement.

Height 17 1/2 in. $1350
Barometer With Clock, unsigned, c. 1900. Beautifully carved oak case in the shape of an anchor with dolphins, an aneroid barometer in the lower section, and an 8-day lever movement in the upper section.
Height 22 in. $500
Richard A. Bourne Co., Inc.

From left to right:
On the left, a Very Rare Miniature Gallery Clock, Brewster & Ingraham, Bristol, Connecticut, c. 1850. Extremely scarce size with laminated and turned mahogany bezel, painted zinc dial inscribed by the maker, 8-day "east-west" movement with overhead suspension, paper label.
Maximum dia. 11 1/4 in. $400
In the middle is a very rare Octagon Wall Clock, Marine Clock Manufacturing Co., New Haven, Connecticut, c. 1847. Mahogany veneered case, painted zinc dial with maker's name and slow/fast indicator, paper label, 30-hour movement with Charles Kirk Patent 2 pallet detached lever escape-

ment and cast iron backplate, very rare and unusual striking mechanism powered by the single main spring.
Height 12 1/2 in. $950
On the right, an Octagon Wall Clock, Litchfield Manufacturing Co., Litchfield, Connecticut, c. 1850. Papier-m>chJ case with inlaid mother-of-pearl and gold decorations, painted zinc dial, 30-hour time-and-strike lever movement with very unusual escapement.
Height 11 1/2 in. $500
Richard A. Bourne Co., Inc.

Three wall clocks, from left to right:
Gallery Clock, Brewster & Ingraham, Bristol, Connecticut, c. 1850. Laminated, turned and gilded bezel with a convex glass, painted wood convex dial, paper label and an 8-day time movement with upside-down suspension and the maker's name inscribed on the front plate.
Dia. 13 3/4 in. $400
Scarce Gallery Clock, Brewster & Ingraham. Bristol, Connecticut, c. 1850. Laminated and turned mahogany bezel, fine convex painted wood dial,

paper label and an 8-day horizontal movement with upside-down pendulum suspension.
Dia. 19 1/4 in. $750
Octagon Wall Clock, E. Ingraham Clock Co., Bristol, Connecticut, c. 1870. Molded oak case, paper dial inscribed by the maker, paper label, 8-day time movement.
Width 14 3/4 in. $300
Richard A. Bourne Co., Inc.

Three assorted wall clocks, from left to right:
Rare Rippled Octagon Wall Clock, Noah Pomeroy, Bristol, Connecticut, c. 1850. This unusual wall clock has a rosewood veneered case with rippled moldings, painted zinc dial, 8-day time movement with lever escapement inscribed by the maker.
Height 10 1/2 in. $600
Octagon Wall Clock, Forestville Hardware & Clock Co., Forestville, Connecticut, c. 1854. A small rosewood veneered octagonal case, dial beautifully inscribed by the maker with seconds indicator, 30-hour movement with lever escapement.
Height 8 in. $695
Gallery Clock, Ansonia Clock Co., Ansonia, Connecticut, c. 1880. In a turned and carved case, paper on zinc dial inscribed, "8 day Time" with the Ansonia trademark, fine large label on the case back, 8-day time movement with lever escapement that runs off two springs onto a common wheel.
Height 9 in. $400
Richard A. Bourne Co., Inc.

Three wall clocks, from left to right:
Rare Octagon Iron Gallery Clock, Noah Pomeroy, Bristol, Connecticut, c. 1865. Painted iron case with mother-of-pearl inlay and painted decorations, paper on zinc dial, 30-hour balance wheel movement marked by the maker.
Height 10 1/2 in. $325
Octagon Wall Clock, Elisha Manross, Forestville, Connecticut, c. 1850. Rosewood veneered case with applied rippled molding and a line of pattern inlay, painted zinc dial and an unusual 8-day time movement with lever escapement that runs off two springs onto a common wheel. The movement is marked "Patented 1849, Elisha Manross, Bristol, Ct. U. S. A. "
Height 12 1/2 in. $495
Octagon Wall Clock, Waterbury Clock Co., Waterbury, Connecticut, c. 1860. Rosewood veneered case, painted zinc dial, paper label, 30-hour time-and-alarm movement with lever escapement.
Height 11 in. $375
Richard A. Bourne Co., Inc.

Rippled Wall Clock, Chauncey Jerome, New Haven, Connecticut, c. 1860. Fine rosewood veneered case with rippled moldings, carved decorations, brass pendulum, painted zinc dial, blue paper maker's label, rare and unusual 8-day time fusee movement.
Height 21 3/4 in. $1450
Richard A. Bourne Co., Inc.

Rare Large Keyhole Clock, Chauncey Jerome, New Haven, Connecticut, c. 1870. Softwood case, black and gold tablet, painted zinc dial, signed on label, 8-day time movement.
Height 18 in. $950
Richard A. Bourne Co., Inc.

Very Rare Octagon Wall Clock with Wagon Spring, Atkins, Whiting & Co., Bristol, Connecticut, c. 1850. The case is rosewood veneered with rippled molding, lower door has its original tablet and opens to a fine label of Atkins, Whiting & Co. ; clock has the rare feature of having a 30-day time-and-strike movement powered by a lever.
Height 25 in. $4600
Richard A. Bourne Co., Inc.

Rare Long Mirror Wall Clock, Joseph Ives, Bristol, Connecticut, c. 1817-1819. Reeded pilaster and scroll, large mirror, original brass finials, upper tablet of classical building, lower tablet is a mourning picture, 8-day short pendulum movement with sculpted cast brass plates and rolling pinions, iron wheels, 1/2 second pendulum.
Height 49 1/2 in. $1800
Museum of Connecticut History

Chapter Three
American Clockmakers

A

Abraham & Albert. Landisburg, Pennsylvania, c. 1832

Adams, Jonas. Rochester, New York. 1834.

Adams, Nathan. Boston, Massachusetts. 1796-1825.

Adams, Thomas F. Baltimore, Maryland. 1804.

Adams, William. Boston, Massachusetts. 1823.

Agar, Edw. Beaver Street, New York. 1761. Sold table clocks.

Alden & Eldridge. Bristol, Connecticut. 1825-1840.

Allebach, Jacobh. Philadelphia, Pennsylvania. 1825-1840.

Almy James. New Bedford, Massachusetts. 1836.

Allen, Jared T. Rochester, New York. 1844.

Alrichs, Jacob. Wilmington, Delaware. 1780-1793.

Alrichs, Jonas. Wilmington, Delaware. 1780-1793.

Alrichs, Jacob and Jonas. Wilmington, Delaware. 1793-1797.

Altmore, Marshall. Philadelphia, Pennsylvania. 1832.

Amant, Fester. Philadelphia, Pennsylvania. 1793.

American Clock Co. New York. 1860.

Anderson, David D. Marietta, Ohio. 1821-1824.

Andrews, L. & F. Bristol, Connecticut. Prior to 1840.

Andrews, N. & T. Meriden, Connecticut. 1832.

Ansonia Brass and Clock Co. Ansonia, Connecticut. 1855.

Ashby, James. Boston, Massachusetts. 1769. Advertised as "Watch-maker and finisher from London, near the British Coffee House in King Street, Boston, Begs leave to inform the Publick that he performs the different Branches of that Business in the Best and Completest Manner at the most Reasonable Rates."

Atkins & Allen Bristol, Connecticut. 1820

Atkins, Eldridge G. Bristol, Connecticut. 1830.

Atkins, Ireneus. Bristol, Connecticut. 1830. Made 30-day brass clocks.

Atkins, Rollin. Bristol, Connecticut. 1826.

Atkins & Porter. Bristol, Connecticut.

Atkins & Son. Bristol, Connecticut. 1870.

Atkins, Whiting & Co. Bristol, Connecticut. 1850-1854.

Atkinson, M. & A. Baltimore, Maryland. 1804.

Atkinson, Peabody. Concord, Massachusetts. 1790. Apprentice to Levi Hutchins.

Austin, Isaac. Philadelphia, Pennsylvania. 1785-1805.

Austin, Orrin. Waterbury, Connecticut. 1820. Established a clock parts factory on Beaver Pond Road.

Avery, . Boston, Massachusetts. 1726. Maker of the clock in the "Old North Church," Boston, from which Paul Revere made arrangements to have the lanterns hung by his friend, Captain John Pulling. Captain Pulling was an ardent patriot, a vestryman of the church and a householder in the nearby vicinity.

Avery, John, Jr. Preston, Connecticut. 1732-1794. A self-taught silversmith and clockmaker, who showed great inventive genius.

B

Babcock & Co. Philadelphia, Pennsylvania. 1832.

Bachelder, Ezra. Danvers, Massachusetts. 1793-1820.

Bacon, John. Bristol, Connecticut. 1830.

Bagnall, Benjamin. Charlestown, Massachusetts. 1712-1740.

Bagnall, Benjamin. Boston, Massachusetts. 1770. Operated a shop at Cornhill, near Town House.

Bagnall, Samuel. Son of the first Benjamin. 1740-1760. Had a shop in Boston.

Bailey & Brothers. Utica, New York. 1847. Motto: "At the Sign of the Big Watch."

Bailey, Putnam. Goshen, Connecticut, c. 1830.

Bailey & Ward. New York, New York. 1832.

Bailey, William. Philadelphia, Pennsylvania. 1832-1846.

Baird Clock Co. Plattsburg, New York. 1875.

Baker, George. Providence, Rhode Island. 1824.

Balch, Benjamin. (Balch & Son) Salem, Massachusetts. 1837.

Balch, Charles H. Newburyport, Massachusetts. Born 1787. In 1811, appointed superintendent of the town clocks.

Balch, Daniel. Newburyport, Massachusetts. 1760-1790.

Balch, Daniel. Son of Daniel 1st. Newburyport, Massachusetts. 1782-1818.

Balch, James. (Balch & Son) Salem, Massachusetts. 1837.

Balch, Moses P. Lowell, Massachusetts. 1832.

Balch, Thomas H. Newburyport, Massachusetts. 1790-1818.

Baldwin, Anthony. Lancaster, Pennsylvania. 1812-1830.

Baldwin, George. Brother of Anthony. Sadsburyville, Pennsylvania. 1808-1832.

Baldwin, Jabez. Boston, Massachusetts 1812.

Baldwin, Jedediah. Hanover, New Hampshire. 1780.

Baldwin, Jedediah. Rochester, New York. 1834.

Baldwin, S. S. & Son. New York, New York. 1832.

Baldwin & Jones. Boston, Massachusetts. 1812.

Banks, Edward P. Portland, Maine. 1834.

Banstein, John. Philadelphia, Pennsylvania. 1791.

Barber, James. Philadelphia, Pennsylvania. 1846.

Barker, B.B. New York, New York. 1790.

Barker, William. Boston, Massachusetts. 1823.

Barnes & Bacon. Bristol, Connecticut. 1840.

Barnes & Bailey. Berlin, Connecticut. 1831.

Barnes, Thomas. Bristol, Connecticut. 1840.

Barnes, Timothy. Litchfield, Connecticut. 1790. Born 1749 in Branford, Connecticut. Parents moved to Litchfield soon after and he spent the rest of his life in Litchfield. He served several weeks in the Revolutionary War. He was a silversmith and a maker of both brass and wood clocks. He died in Litchfield on October 11, 1825.

Barrows, James M. Tolland, Connecticut. 1832.

Barry, Standish. Baltimore, Maryland. 1804.

Bartholomew, E. & G. Bristol, Connecticut. Circa 1820.

Barton, Benjamin. Alexandria, D. C. 1832. Advertised as "Clock and Watch Maker, Keeps for sale a general assortment of Clocks, Watches, etc. South side of King Street between Fairfax and Royal."

Barton, O.H. Fort Edwards, New York. 1870.

Bassett, N. B. Albany, New York. 1813.

Bateson, John. Boston, Massachusetts. 1720. Upon his death in 1727, it was found that he had left in his shop an eight-day clock movement valued at 25 pounds 10 shillings and a silver repeating watch valued at 90 pounds.

Batterson, James. Boston, Massachusetts. 1707-1730

Battle, A. B. Utica, New York. 1847.

Baugh, Valentine. Abingdon, Virginia. 1820-1830.

Bauer, John N. New York, New York. 1832.

Bayley, Calvin. Hingham, Massachusetts. 1800.

Bayley, John. Hanover, Massachusetts. 1770-1815.

Bayley, John. Son of John, Hingham, Massachusetts. 1814-1820.

Bayley Joseph. Hingham, Massachusetts. 1808.

Bayley, Simeone C. Philadelphia, Pennsylvania. 1794.

Beard, Duncan. Appoquinemonk, Delaware. 1755-1797.

Beach & Hubbard. Bristol, Connecticut. 1869. Marine Clocks.

Belk, William. Philadelphia, Pennsylvania. 1796.

Belknap, Ebenezer. Boston, Massachusetts. 1823.

Bell, James. New York, New York. 1804.

Bell, John. New York, New York. In 1734, he was advertising 8-day clocks with Japan Cases.

Benedict & Burnham Co. Waterbury, Connecticut. 1850-1855. Chauncey Jerome of New Haven and Bristol, Connecticut was induced to enter into business with this company, and upon its failure, he suffered complete ruin.

Benedict, S. W. 30 Wall St., New York, New York. 1829. Advertised as paying "Particular attention to the repairs of watches and clocks."

Benjamin, Barzillai. New Haven, Connecticut. 1823. Advertising read "Gold and Silver watches, Duplex or vertical movements, warrented for one year."

Bennett, Alfred. Portsmouth, New Hampshire. c. 1868.

Berwick, Abner. Berwick, Maine. 1820.

Bevans, William. Norristown, Pennsylvania. 1816.

Bigger & Clarke. Baltimore, Maryland. 1784.

Bigger, Gilbert. Baltimore, Maryland. 1799.

Bill, Joseph Rogers. Middletown, Connecticut. 1840.

Billow, Charles & Co. Boston, Massachusetts. 1796.

Birdsey, E. C. & Co. Meriden, Connecticut. 1831. "Manufacturers of improved clocks."

Birge, Gilbert & Co. Bristol, Connecticut. 1835.

Birge, John. Bristol, Connecticut. 1830-1837. Initially a wagon builder; later purchased the patent for the rolling pinion eight-day brass clock. Obtained a factory, manufactured clocks and marketed them by means of traveling salesmen, the "peddlers" of the day. Farmed and made clocks until shortly before his death.

Birge & Fuller. Bristol, Connecticut. 1845.

Birge, Mallory & Co. Bristol, Connecticut. 1830.

Birge, Peck & Co. Bristol, Connecticut. 1830-1856.

Bisbee, J. Brunswick, Maine. 1798-1825.

Bishop & Bradley. Plymouth, Connecticut. 1825-1830. (James Bishop and L. B. Bradley.)

Bissell, David. East Windsor, Connecticut. 1832. Advertised as "Watch and Clock Maker and Dentist."

Bixler, Christian. Easton, Pennsylvania. 1785-1830.

Blakeslee, Jeremiah. Plymouth, Connecticut. 1841-1849 Associated with Myles Morse.

Blakeslee, Marvin & Edward. Heatherville, Connecticut. (near Plymouth). 1832.

Blakesley (or Blakeslee), Milo. Plymouth, Connecticut. Employed by Eli Terry Jr. about 1824, and later became Terry's partner.

Blasdell, David. Amesbury, Massachusetts. 1741.

Blasdell, Isaac. Chester, New Hampshire. 1762-1791. Son of David Blasdell, Amesbury. Isaac Blasdell's clocks were probably the first clocks made in New Hampshire.

Blasdell, Richard (possibly the son of Isaac). Chester, New Hampshire. About 1788.

Boardman, Chauncey. Bristol, Connecticut. 1813. Began making clocks about 1813 in North Forestville, Connecticut. Started the manufacture of brass clocks about 1838 and continued until his business failed in 1850.

Boardman & Dunbar. Bristol, Connecticut. 1811.

Boardman & Hubbell. Forestville, Connecticut. 1867.

Boardman & Wells (Chauncey Boardman and Joseph A. Wells). Bristol, Connecticut. 1815. Shortly after 1820, these partners built a factory in North Forestville, Connecticut, undoubtedly one of the most important at that time.

Bode, William. Philadelphia, Pennsylvania. 1796.

Bogardus, Everadus. New York, New York. 1698.

Bond, William. Boston, Massachusetts. 1800-1810

Bonfanti, Joseph. 305 Broadway, New York, New York. 1835. A specialist in French and German clocks.

Bonnaud, ——. Philadelphia, Pennsylvania. 1799

Boss & Peterman. Rochester, New York. 1841. Advertised as follows: "We strive to Excel. Dealers in Watches and Jewelry. Try us before pursuing elsewhere. We feel warranted in saying that all watch and clock work entrusted to our care will be executed better than at any other establishment in this city."

Botsford, S. N. Hamden (Whitneyville), Connecticut. 1856.

Boughell, Joseph. New York, New York. 1787.

Bower, Michael. Philadelphia, Pennsylvania. 1790-1800.

Bowman, Joseph. Lancaster, Pennsylvania. 1821-1844.

Bowne, Samuel. New York, New York. 1751. Advertised "Japanned and Walnut cased Clocks."

Brace, Rodney. North Bridgewater (now Brockton), Massachusetts. 1830. In the early 1800s, Rodney Brace, formerly of Torrington, Connecticut, began manufacturing small wooden clocks with Isaac Parkard. They were the first to introduce small clocks, which were shipped by wagon to all parts of the country.

Bradley & Barnes. Boston, Massachusetts. 1856.

Bradley & Hubbard. Meriden, Connecticut. 1854.

Bradley, Nelson. Plymouth, Connecticut. 1840.

Bradley, Richard. Hartford, Connecticut. 1825-1839. Watch repairer.

Bradley, Z. & Son. New Haven, Connecticut. 1840.

Brandegee, Elishama. Berlin, Connecticut. 1832. Advertised as "Manufacturer of Cotton Thread, Clocks of all descriptions, and Dealer in American Goods."

Brandt & Mathey. Philadelphia, Pennsylvania. 1799.

Brandt, Brown & Lewis. Philadelphia, Pennsylvania. 1795.

Brasher, Abraham. New York, New York. 1757.

Brasier, Amamble. Philadelphia, Pennsylvania. 1795-1820.

Brastow, Adison & Co. Lowell, Massachusetts. 1832.

Brearley, James. Philadelphia, Pennsylvania. 1793-1811.

Breckenridge, J. M. Meriden and New Haven, Connecticut. (born 1809, died 1896). His obituary read: "Mr. Breckenridge, who died at the age of 87, had been a clockmaker all of his life, and his death removed the last of the original Connecticut clockmakers. He learned his trade when 19, and lived through the wonderful growth of the clockmaking business, saw wood clocks crowd out cast brass, and sheet brass take the place of wood. Although a clockmaker for so many years Mr. Breckenridge never acquired a fortune, as so many of his contemporaries did, and was at his bench in the shop of the New Haven Clock Co. till within a few months of his death. His last work was making dies for clock hands, which requires special skill. During his long career, he made many improvements in the tools for making clocks, among them the punch-box or pick-off die, which punches pivot holes in the brass frame. When a young man he invented the wire clock bell or gong, and fifty years ago these bells were used on nearly all the large clocks, but few are employed now. In 1850, Mr. Breckenridge went into the powder flask business in Springfield, Massachusetts, but finding it unprofitable returned to his clockmaking work, which he never again gave up till a few months before his death."

Brewer, Isaac. Philadelphia, Pennsylvania. 1813.

Brewer, William. Philadelphia, Pennsylvania. 1785-1791.

Brewster, Elisha C. Bristol, Connecticut. 1833-1862. Born 1791, died 1880. Sold clocks throughout the South for Thomas Barnes of Bristol. Was a clock worker in the town of Plainville for several years, then became involved in the painting of clocks, dials and glass in the city of Bristol, Connecticut. In 1833, he bought the business and factory of Charles Kirk, keeping him on as manager until 1838. With Shaylor Ives, he invented a new spring for clocks and manufactured the first spring clocks made in this country. In 1843, a partnership was formed with Elias and Andrew Ingraham, which lasted until 1848 when he bought them out; he later brought in William Day and Augustus Norton, buying them out at a later date, and continuing the business by himself until 1862, at which time he retired. He operated a branch store in London, England for about 24 years, 20 of which saw the management of the store by his son, N.L. Brewster.

Brewster & Ingraham. Bristol, Connecticut. 1843-1848.

Brinsmaid & Bros. Burlington, Vermont. c. 1840.

Brinckerhoff, Dirck. Dock Street, New York, New York. 1750. "At the sign of the Golden Duck."

Bronson, I.W. Buffalo , New York. 1825-1830 "Improved brass clocks."

Brooks, B. F. Utica, New York. 1847.

Brown, David. Providence, Rhode Island. 1834-1850.

Brown, Gawen. Boston, Massachusetts. 1750-1776.

Brown, J. C. Bristol, Connecticut. 1827-1855. In 1835 a factory for the manufacture of clocks was built in Bristol, Connecticut. by J. C. Brown, William Hills, Jared Goodrich, Lora Waters and Chauncey Pomeroy, and was occupied by the Welch Co. when Mr. Brown later bought out the other investors and built the J. C. Brown shop in 1853. When this business failed the shop was acquired by Mr. Welch, and later became the E. N. Welch Mfg. Co.

Brown, Joseph R. Providence, Rhode Island. 1849.

Brown & Kirby. New Haven, Connecticut. 1840.

Brown, Laurent. Rochester, New York. 1841. "Watch and Clock Establishment."

Bryant, Thomas. Rochester, New York. 18—.

Bullard, Charles. Boston, Massachusetts, b. 1794; d. 1871. Apprentice of and successor to the English artist who specialized in painted glass fronts and dials for Simon Willard.

Burdick, M.H. Bristol, Connecticut. 1849.

Burkelow, Samuel. Philadelphia, Pennsylvania. 1791-1799.

Burnap, Daniel. East Windsor and Andover, Connecticut. b. 1759; d. 1838. Learned the clockmaking trade from Thomas Harland, Norwich, Connecticut. Settled in East Windsor about 1780 (some say 1776) where he carried on an extensive trade in clocks; one of his first apprentices was Eli Terry, who later became one of the foremost clockmakers in the country. Shortly before 1800, he moved back to the Coventry area, where his business remained until his death in 1838. Burnap was also a skilled silversmith. His clocks were noted for their tall cases, brass works, silvered dials, and fine engraving. Although he probably made some wooden clocks, his fame rests upon the handsome eight-day clocks that were his specialty.

Burr, C.A. Rochester, New York. 1841. Advertisement: "Wholesale and retail Dealer in Watches, Clocks, Jewelry, etc. Has on sale Gold, Silver, Duplex, Anchor, Independent sec'ds, Patent Lever, Lepine and Vertical Watches, French, Mantel, Wood, Brass 30 hour and 8 day clocks."

Burr, Ezekial & William. Providence, Rhode Island. 1792.

Burritt, Joseph. Ithaca New York. 1831.

Burwell & Carter. Bristol, Connecticut. 1860.

Bush, George. Easton, Pennsylvania. 1812-1837.

Butler, N. Utica, New York. 1803.

Byington & Co., L. Bristol, Connecticut. 1849.

C

Cain, C. W. New York, New York. 1836.

Cairns, John. Providence, Rhode Island. 1784.

Cairns, John, 2nd. Providence, Rhode Island. 1840-1853.

Calendar Clock Co., The,. Glastonbury, Connecticut. 1856.

Camp, Ephraim. Salem Bridge, Connecticut. 1830.

Camp, Hiram. New Haven, Connecticut. 1829-1893. Nephew of Chauncey Jerome of Bristol, Connecticut. He started his clockmaking career when he went to work for his uncle in 1829 at the Jerome Mfg. Co.; in 1845, he accompanied the business when it relocated to New Haven. In 1855, when the Jerome Mfg. Co. failed, it was succeeded by the New Haven Clock Co., where Hiram Camp spent forty years as its president.

Campbell, Charles. Philadelphia, Pennsylvania. 1795-1799.

Campbell, R. A. Baltimore, Maryland. 1832.

Campbell, William. Carlisle, Pennsylvania (1765), Philadelphia, Pennsylvania (1799).

Canby, Charles. Wilmington, Delaware. 1815-1850.

Capper, Michael. Philadelphia, Pennsylvania. 1785-1799.

Carrell, John and Daniel. Philadelphia, Pennsylvania. 1791-1793.

Carter, L.F. & W. 1860s.

Carver, Jacob. Philadelphia, Pennsylvania. 1785-1799.

Cary, James, Jr. Brunswick, Maine. 1806-1850. Apprenticed to Robert Eastman in 1805, made a partner 1806-1809. In 1830, Aaron L. Dennison was apprenticed to Cary.

Case & Birge (Erastus and Harvey Case, with John Birge). Bristol, Connecticut. 1830-1837.

Case, Dyer, Wadsworth & Co. Augusta, Georgia. 1835. Assembled and sold subcontracted parts. Cases and movements were made for them by Seth Thomas.

Case, Erastus (brother of Harvey). Bristol, Connecticut. 1830-1837. Made eight-day clocks with brass works with John Birge. Retailed mainly in the South, with annual output about 4,000 units.

Case, Willard & Co. Bristol, Connecticut. 1835.

Case, Harvey. Bristol, Connecticut. 1830-1837.

Case & Robinson. Bristol, Connecticut. 1856.

Casten, Stephen, & Co. Philadelphia, Pennsylvania. 1819.

Cate, Col. Simeon. Sanbornton, New Hampshire. 18?

Chadwick, Joseph. Boscawen, New Hampshire. 1810-1831.

Chandlee, Benjamin. Baltimore, Maryland. 1817.

Chandlee, John. Wilmington, Delaware. 1795-1810.

Chandler, Abiel. Concord, New Hampshire. 1829-1858. Son and successor to Maj. Timothy Chandler. As an adjunct to clockmaking, he also made mathematical instruments.

Chandler, Major Timothy. Concord, New Hampshire. 1785-1840. In 1783, he made his way from Pomfret, Connecticut to Concord, New Hampshire where he set up a clockmaking shop, hiring an apprentice by name of Cummings from Simon Willard.

Chapin, Aaron & Son. Hartford, Connecticut. 1825-1238.

Chase, Timothy. Belfast, Maine. 1826-1840.

Chaudron, P. Philadelphia, Pennsylvania. 1799.

Chaudron, S. & Co. Philadelphia, Pennsylvania. 1811.

Cheeny, J. East Hartford, Connecticut. 1790

Cheney, Asahel. Northfield, Massachusetts. 1790. Eldest son of Benjamin Cheney, he was born about 1758. Learned the trade of clockmaking at his father's shop in East Hartford, Connecticut, later moving to Northfield, Mass where he carried on an extensive business in the manufacture of eight day clocks.

Cheney, Benjamin (Brother of Timothy). East Hartford, Connecticut (Section now known as Manchester). 1745-1780. Numbered among the first clockmakers in New England. Clocks were noted for having tall, carved cherry wood cases, and wooden works. John Fitch, inventor of the steamboat, was an apprentice in their shop.

Cheney, Elisha (son of Benjamin). Berlin, Connecticut. 1800-1835. Received his mechanical training from his father, making both brass and wooden tall clocks in his Berlin shop. Was in partnership for a short time with his brother-in-law, Simeon North, making pistols. In 1801, bought a shop on the Berlin-Middletown line and made wooden tall and mantel clocks.

Cheney, Olcott (son of Elisha). Middletown and Berlin, Connecticut. About 1820-1850. After working with his father until 1835, he bought his father out, carrying on the business until 1850. Although he resided in Berlin, his shop was just over the line in Middletown, thus Cheney clocks were marked "Middletown."

Cheney, Russell (son of Benjamin and brother of Elisha). East Hartford, Connecticut and Thetford, Vermont. Became a skillful workman at the family trade of clockmaking, but there are no records to indicate how long he continued in the business.

Cheney, Timothy (brother of Benjamin). East Hartford, Connecticut. About 1776-1795. A skilled clockmaker, watch repairer, blacksmith, joiner and silversmith. He probably learned the clockmaking trade from his brother, and made both brass and wooden clocks. There is nothing to indicate that Timothy and Benjamin were ever in business together; while they both lived and worked in East Hartford, it appears they did so independent of each other.

Chester, George. New York, New York. 1757.

Child, John. Philadelphia, Pennsylvania. 1813-1835.

Child, True W. Boston, Massachusetts. 1823.

Chollet, John B. Philadelphia, Pennsylvania. 1819.

Church, Joseph. Hartford, Connecticut. 1825-1838.

Church, Lorenzo. Hartford, Connecticut. 1846.

Clagget, H. Newport, Rhode Island. 1726-1740

Clagget, Thomas (brother of William). Newport, Rhode Island. 1720-1749.

Clagget, William (brother of Thomas). Newport, Rhode Island. 1720-1749.

Clark, Benjamin. Wilmington, Delaware. 1837-1850.

Clark, Benjamin. Philadelphia, Pennsylvania. 1790-1819.

Clark, Benjamin & Ellis. Philadelphia, Pennsylvania. 1813.

Clark, Daniel. Waterbury, Connecticut. 1814-1820. In company with Zena Cook and William Porter, Clark owned a clock factory in Waterbury, Connecticut.

Clark, Edward. Philadelphia, Pennsylvania. 1797.

Clark, Ellis. Philadelphia, Pennsylvania. 1811-1845.

Clark, Ephraim. Philadelphia, Pennsylvania. 1780-1810.

Clark, Herman. Plymouth Hollow (now Thomaston), Connecticut. 1807.

Clark, Jesse, W & C. Philadelphia, Pennsylvania. 1811.

Clark, Joseph. New York, New York. 1768. His advertisement read: "Some exceedingly good 8 day clocks in very neat mahogany cases." He later moved to Danbury, Connecticut, where he maintained his shop until 1811, when he returned to New York state. His later years were spent in Alabama where he died in 1821.

Clark, Sylvester. Salem Bridge (now Naugatuck), Connecticut. 1830.

Clarke, Charles. Philadelphia, Pennsylvania. 1806-1811

Clarke, George G. Providence, Rhode Island. 1824.

Clarke, Gilbert & Co. Winsted, Connecticut. 1842.

Clarke, Lucius. Winsted, Connecticut. 1841. In 1841, Clarke purchased the business of Riley Whiting from Whiting's estate, and associated himself with William L. Gilbert as Clarke, Gilbert & Co.

Clarke, John. New York, New York. 1770-1790.

Clarke, John. Philadelphia, Pennsylvania. 1799.

Claton, C. Philadelphia, Pennsylvania. -?-

Clements, Moses. New York, New York. 1749.

Cleveland, Benjamin Norton. Newark, New Jersey, b. 1767; d. 1837.

Cleveland, William, Salem, Massachusetts. 1780. Apprenticed to Thomas Harland. Was a shipowner, merchant and watchmaker.

Cleveland, William (nephew of other William). Worthington, Massachusetts, Salem, Massachusetts, and Norwich, Connecticut. Grew up in Norwich, where he learned the trade of silversmith, watch and clockmaker from Thomas Harland. At the age of 21, he set up shop in New London with John Proctor Trott; this partnership dissolved in 1796. After marrying, he established another shop in Worthington, Massachusetts, later moving the business to Salem, Mass for a few years. He then went to New York, returning to Norwich in 1812, where he lived until his death in 1837. He was the grandfather of President Grover Cleveland.

Coe, Russell. Meriden, Connecticut. 1856.

Cole, James C. Rochester, New York. 1812-?-. Apprenticeship with Edward S. Moulton, after which he established the same business, adding watchmaker and silversmith.

Cole, Shubael (son of James C. Cole). Great Falls (now Somersworth) New Hampshire. 18?

Conant, Elias. Bridgewater, Massachusetts. 1776-1812.

Conant, Elias. Lynn, Massachusetts. 1812-1815.

Conant, W.S. New York City. about 1820.

Conant & Sperry. 1840.

Conrad, O. Philadelphia, Pennsylvania. 1846.

Cook, E. Rochester, New York. 1824.

Cook, Zenas. Waterbury, Connecticut. 1811-1820. Partner of Daniel Clark and William Porter in ownership of clock factory built on Great Brook, Waterbury in 1814.

Cooper, T. Olneyville, Rhode Island. 1849.

Corey, P. Providence, Rhode Island. 1849.

Corliss, James. Weare, New Hampshire. 1800.

Cornell, Walter. Newport, Rhode Island. -?-

Couper, Robert. Philadelphia, Pennsylvania. -?-

Cox & Clark. New York, New York. 1832. "Importers and Dealers in Lamps, etc.; also French China and Mantel Clocks, Silver Ware, etc."

Cozens, Josiah B. Philadelphia, Pennsylvania. 1819.

Cranch, Richard. Boston and Braintree, Massachusetts. 1771-1789. Prior to the Revolutionary War, he carried on his business near the Mill Bridge in Boston, where he sold all kinds of watch and clockmaker's tools.

Crane, Aaron. Newark, New Jersey. c. 1851.

Crane, Simeon. Canton, Massachusetts. -?-

Crehore, Charles Crane. Boston, Massachusetts, b. 1793; d. 1879. Made clock cases for Simon Willard Sr. and Jr., as well as for Benjamin F. Willard and many other clockmakers.

Crichet, James. Candis, New Hampshire. About 1800.

Crocker, Orsamus. East Meriden (or Bangall), Connecticut. About 1831. Built a factory for clockmaking, but it was a failure.

Crosby & Vosburgh. New York. c. 1860. Successors to Jerome Manufacturing Co.

Cross, James. Rochester, New Hampshire. 18?.

Crow, George. Wilmington, Delaware. 1740-1770.

Crow, John. Wilmington, Delaware. 1770-1798.

Crow, Thomas. Wilmington, Delaware. 1770-1824.

Crow, Thomas. Philadelphia, Pennsylvania. 1795.

Crow, Thompson. Wilmington, Delaware. -?-

Crowley, John. Philadelphia, Pennsylvania. 1813.

Crowther, William. New York, New York. 1820.

Cummens, William. Roxbury, Massachusetts. 1788-1834. Apprenticed to Simon Willard; then engaged in the business of clockmaking, where he was noted for long case, shelf clocks and timepieces. Had a son William, who assisted him. His clocks were usually marked, "Warranted by William Cummens, Roxbury."

Cure, Lewis. Brooklyn, New York. 1832.

Currier, Edmund. Salem, Massachusetts. 1837. "Watches, Timepieces, Gallery Clocks and Regulators."

Curtiss & Clark. Plymouth, Connecticut. 1824.

Curtis & Dunning. In the Red Lion Inn, Stockbridge, Massachusetts, the handsomest clock on display, banjo shaped and entirely gilt is marked "Curtis" on the dial. This name is also on an unusually elegant banjo clock owned at one time by Mrs. S. C. McKown, Rochester, N. Y.

Curtis, Lemuel. Concord, Mass and Burlington, Vermont. 1814-1857. In 1816, Curtis took out a patent on an improvement on the Willard timepiece. In 1818, he moved to Burlington, Vermont, where he remained until his death in 1857. Clocks made by Curtis were patterned after the Willard models, but he used more ornamentation and more pleasing proportions. One outstanding feature of Curtis' clocks is the circular pendulum box.

Curtis, Lewis. Farmington, Connecticut. b. 1774; d. 1845. In 1795, aided by his father, he opened a shop on Main Street, advertising clockmaking and silversmithing. In 1820 He moved to St. Charles, Mississippi and subsequently went to Hazel Green, Wisconsin, where he died in 1845.

Curtis, Solomon. Philadelphia, Pennsylvania. 1793.

Custer, Jacob D. Norristown, Pennsylvania, b. 1805; d. 1872. In 1832, he began the manufacture of "grandfather clocks." In 1842, he started manufacturing the clockwork mechanism that propelled the rotating lights in lighthouses.

D

Daft, Thomas. New York, New York. 1786.

Daft, Thomas. Philadelphia, Pennsylvania. -?-

Daggett, T. Providence, Rhode Island. 1849.

Dalziel, John. New York, New York. 1798.

Dana, George. Providence, Rhode Island. 1805. In company with Thomas Whitaker, he bought out the business of Nehemiah Dodge.

Dana, Payton. Providence, Rhode Island. 1849.

Dana, Peton and Nathaniel. Providence, Rhode Island. 1800.

Darrow, Elijah. Bristol, Connecticut. 1822-1830. He joined Chauncey and Noble Jerome in 1824.

Davidson, Barzillai. Norwich, Connecticut. 1775. An accomplished worker in gold and silver, he sold jewelry and timepieces as well.

Davidson, Barzillai. New Haven, Connecticut. 1825. The clock with wooden works in the New Haven Meeting House was made by Davidson in 1825.

Davis & Babbitt. Providence, Rhode Island. 1810.

Davis, David P. Roxbury, Massachusetts. 1847-1856. Partner of Edward Howard.

Davis, John. New Holland Patent, Pennsylvania. 1802-1805.

Davis, Peter. Jaffrey, New Hampshire. -?-

Davis, Samuel. Pittsburgh, Pennsylvania. 1815.

Davis, William. Boston, Massachusetts. 1683. Emigrated from England to pursue the trade of clockmaker; David Edwards became surety for the entire Davis family in order that they would not become charges of the town.

Dawson, Jonas, Philadelphia, Pennsylvania. 1813.

DeForest & Co. Salem Bridge, New York. 1832. They advertised as follows: "Watches and clocks and buttons of all kinds are manufactured."

Delaplaine, James. New York, New York. 1786-1800

Deloste, Francis. Baltimore, Maryland. 1817.

Demilt, Thomas, Philadelphia, Pennsylvania. 19th century.

Demilt, Thomas. New York, New York. 1798-1818.

Demilt, Thomas & Benjamin. New York, New York. 1802-1818.

Dennison, Aaron L. Roxbury and Waltham, Massachusetts and Birmingham, England. 1850-1895. Considered a pioneer in American watchmaking. Dennison and Edward Howard started a watch factory at Roxbury in 1850, which they moved to Waltham in 1854. When the factory was sold in 1857, Dennison stayed on as superintendent for an unknown period. After several changes, this became known as the Waltham Watch Co. Dennison later settled in Birmingham, England, where he designed machinery for the making of watch cases.

Derby, Charles. Salem, Massachusetts. 1846-1850.

Derby, John. New York, New York. 1816.

De Riemer & Mead. Ithaca, New York. 1831.

De Saules & Co. New York, New York. 1832.

Deverell, John. Boston, Massachusetts. 1789-1803.

Dewey, I. Chelsea, Vermont. 1830.

Dexter, Joseph W. Providence, Rhode Island. 1824.

De Young, Meichel. Baltimore, Maryland. 1832.

Dix, Joseph. Philadelphia, Pennsylvania. -?-

Dobbs, Henry M. New York, New York. 1794-1802.

Dodge, Ezra W. Providence, Rhode Island. 1824.

Dodge, George. Salem, Massachusetts. 1837.

Dodge, Nehemiah. Providence, Rhode Island. 1794-1824. In 1799, Dodge was associated with Stephen Williams for a short time, while later his partner was Gen. Josiah Whitaker. Upon retiring, the business was sold to George Dana and Thomas Whitaker.

Dodge, Seril. Providence, Rhode Island. 1788. He served his apprenticeship with Thomas Harland, and was also known as a gold and silversmith.

Dominick, Friedrich. Philadelphia, Pennsylvania. -?-

Doolittle, Enos. New Haven, Connecticut, b. 1751; d. 1806. Apprenticed to his uncle, Isaac Doolittle in New Haven; at the age of 21, made his first clock and moved to Hartford, Connecticut. He continued in the trade, and in 1787 got into the brass foundry business with Jesse Goodyear of Hamden. He remained a "Jack at all Trades" until his death in 1806.

Doolittle, Isaac. New Haven, Connecticut, b. 1721; d. 1800. Apprentice of Macock Ward, Wallingford, Connecticut, moving to New Haven and opening his own shop in 1742. His business pursuits over many years included armorer for the Connecticut Fourth Regiment, powder mill operator, foundry owner and purveyor of clocks, watches, chocolate, surveying instruments and caster of bells.

Doty, John F. Albany, New York. 1813.

Douglas, John. New Haven, Connecticut. 1800-1820.

Douty, Henry. Philadelphia, Pennsylvania. -?-

Dowdney, Burrows. Philadelphia, Pennsylvania. -?-

Dowling, G. R. & B., Co. Newark, New Jersey. 1832.

Downs, Ephraim. Bristol, Connecticut. 1811-1843. Started his clockmaking with Lemuel Harrison of Waterbury in 1811, later going to Cincinnati, Ohio to make clocks for Lumas Watson during the period 1816-1821. Settled in Greystone working on clocks with Seth Thomas, Eli Terry and Seth Hoadley, Terry's brother-in-law. Opened his own business, but in 1825 moved to Bristol, where from "Down's Mill" clocks were sent to New York, New Jersey, Pennsylvania, Ohio, Missouri, Mississippi, and elsewhere. A favorite design of his was the "looking glass" clock. Of all the Bristol clockmakers, he alone was unaffected by the business recession of 1837. Retired in 1843.

Droz, Charles A. Philadelphia, Pennsylvania. 1813.

Droz, Humbert. Philadelphia, Pennsylvania. 1797.

Droz, Humbert A.L. Philadelphia, Pennsylvania. 1811.

Droz & Sons. Philadelphia, Pennsylvania. 1813.

Drysdale, William. Philadelphia, Pennsylvania. 1819-1851.

Dubois & Folmar. New York, New York. 1816.

Ducommun, A.L. Philadelphia, Pennsylvania. 1797.

Dudley, Benjamin. Newport, Rhode Island. 1840.

Duffield, Edward. Philadelphia, Pa, 1741-1747. Lower Dublin, Pa, 1747-1801. A personal friend of Benjamin Franklin, he started his trade of clock and watch maker in Philadelphia, later moving to Lower Dublin where he died in 1801.

Duffield, Edward. West Whiteland Township, Chester County, Pennsylvania. -?-

Dunbar, Butler. Bristol, Connecticut. 1810-1830. In 1810, he was making clocks with Dr. Titus Merriman; he later went to Springville, Pennsylvania.

Dunbar, Jacobs & Warner. Bristol, Connecticut. 1849.

Dunbar & Merriman. Bristol, Connecticut. 1810 -?-

Dunheim, Andrew. New York, New York. 1775.

Dunlap, Archibald. New York, New York. 1802.

Dunning & Crissey. Rochester, New York. 1847.

Dunning, J.L. Connected with Lemuel Curtis.

Dupuy, John. Philadelphia, Pennsylvania. 1770.

Dupuy, Odran. Philadelphia, Pennsylvania. 1735.

Durgin, F. Andover, New Hampshire. -?-

Dutch, Stephen, Jr. Boston, Massachusetts. 1800-1810.

Dutton, David. Mount Vernon, New Hampshire. -?-

Dyer, Warren. Lowell, Massachusetts. 1831. Advertised as follows: "Clocks and Timepieces of brass, eight-day movements, set up and warrented correct timekeepers. Prices from 9 to 25 dollars."

Dyer, Joseph. Concord, Massachusetts. 1815-1829. Dyer was a journeyman with Lemuel Curtis, and, when Curtis moved to Burlington, Vermont, Dyer took over and continued the business in Concord; he later moved to Middlebury, Vermont.

E

Easterley, John. New Holland, Pennsylvania. 1825-1840.

Eastman, Abel B. Belfast, Maine. 1806-1821. Considered the earliest clockmaker in Belfast, having come from Concord, New Hampshire in 1806.

Eastman, Abel B. Haverhill, Massachusetts. 1816-1821.

Eastman & Cary. Brunswick, Maine. 1806-1809. Eastman sold the business to Cary in 1809

Eastman, Robert. Brunswick, Maine. 1805-1808. James Cary Jr. was his apprentice; Cary was taken in partnership in 1806 and the business was sold to Cary in 1809.

Eaton, John H. Boston, Massachusetts. 1823.

Eberman, John. Lancaster, Pennsylvania. 1780-1820.

Edson, Jonah. Bridgewater, Massachusetts. 1815-1830.

Edwards, Abraham. Ashby, Massachusetts. 1794-1840. A self-taught clockmaker.

Edwards, Samuel. Gorham, Maine. 1808-? Came to Maine from Ashby, Massachusetts in 1808 and carried on the manufacture of wooden clocks for many years.

Eliot, William. Baltimore, Maryland. 1799.

Elliot, Hazen. Lowell, Massachusetts. 1832.

Elsworth, David. Windsor, Connecticut. 1780-1800. Brother of Oliver Elsworth, Chief Justice of the United States. Apprenticed to either Seth or Benjamin Youngs of Windsor. Began business as a clockmaker, watch repairer and dentist in Windsor in 1763. He also made muskets for the army during the Revolution. Died in 1821 at the age of seventy-eight.

Elvins, William. Baltimore, Maryland. 1799.

Embree, Effingham. New York, New York. 1785-1794.

Emery, Jesse. Weare, New Hampshire. 1800. First clockmaker in the town of Weare.

Ent, Johann. Philadelphia, Pennsylvania. -?-

Ent, John. New York, New York. 1758.

Essex, Joseph. Boston, mass. 1712.

Evans, David. Baltimore, Maryland. 1770-1773. Advertised "At the Sign of the Arch, Dial and Watch, Gay Street."

Evans, Thomas. New York, New York. 1766.

Evans, William M. Philadelphia, Pennsylvania. 1813-1819 or possibly longer.

Eyre, Johann. Philadelphia, Pennsylvania.

F

Fahrenbach, Pius. Boston, Massachusetts. 1856.

Fales, G.S. New Bedford, Massachusetts. 1827.

Fales, James. New Bedford, Massachusetts. 1810-1820.

Fales, James Jr. New Bedford, Massachusetts. 1836.

Farnham, S.S. Oxford, New York. 1842.

Farnum, Henry & Rufus. Boston, Massachusetts. 1780. Apprentices of Thomas Harland.

Farr, John C. Philadelphia, Pennsylvania. 1832.

Favre, John James. Philadelphia, Pennsylvania. -?-

Fellows, James K. Lowell, Massachusetts. 1832.

Fellows, Read & Olcott. New York, New York. 1829.

Fellows, Storm & Cargill. New York, New York. 1832.

Ferris, Benjamin, C & W. Philadelphia, Pennsylvania. 1811.

Ferris, Tiba. Wilmington, Delaware. 1812-1850.

Fessler, John. Fredericktown, Maryland. 1782-1820.

Feton, J. Philadelphia, Pennsylvania. 1828-1840.

Field, Peter. New York, New York. 1802.

Field, Peter Jr. New York, New York. 1802-1825.

Fiffe, H. No address given. Maker of banjo clocks.

Filber, John. Lancaster, Pennsylvania. 1810-1825.

Fish, Isaac. Utica, New York. 1846.

Fisk, William. Boston, Massachusetts, b. 1770; d. 1844. Made clock cases for different clockmakers, including Aaron Willard and Simon Willard, for whom he made all the clock cases during the period 1800-1838.

Fite, John. Baltimore, Maryland. 1817.

Fitz, ——. Portsmouth, New Hampshire. about 1769.

Fix, Joseph. Reading, Pennsylvania. 1820-1840.

Fletcher, Charles. Philadelphia, Pennsylvania. 1832.

Fletcher, Thomas. Philadelphia, Pennsylvania. 1832. "Manufactory of Jewelry and Silver Ware, and Furnishing Warehouse. Extensive assortment of Watches...Mantel Clocks..."

Fling, Daniel. Philadelphia, Pennsylvania. 1811.

Flower, Henry. Philadelphia, Pennsylvania. 1753.

Folmar, Andrew. New York, New York. 1810.

Foot, Charles J. Bristol, Connecticut. 1856.

Foote, William. Middletown & East Haddam, Connecticut. b. 1772; d. ? Foote went into partnership with Samuel Canfield, a silversmith, in 1792 at Middletown. The partnership was dissolved in 1796, and Foote moved to East Haddam, opening a shop near the landing where he continued to make clocks. At a later point, he moved to Michigan, where he died some time after 1836.

Forbes, John. Hartford, Connecticut & Philadelphia. Forbes, a clock and watch maker from Philadelphia, Pennsylvania settled in Hartford, Connecticut in 1770.

Forbes, Wells. Bristol, New Hampshire. About 1840.

Forestville Hardware & Clock Co. Forestville, Connecticut. c. 1854.

Forestville Manufacturing Co. Bristol, Connecticut. About 1830.

Foster, John C. Portland, Maine. 1834.

Foster, Nathaniel. Newburyport, Massachusetts. 1818-1828. In 1818, Foster opened a shop on State Street where he carried on "the clock and watch making business in all its branches." He was appointed in charge of all the town clocks from 1818 until 1828, and perhaps even longer.

Fowell, J. & N. Boston, Massachusetts. 1800-1810.

Francis, Basil & Alexander. Vuille, Baltimore, Maryland. 1766.

Frary, Obadiah. Southampton, Massachusetts. 1745-1775. Made a number of brass clocks for local families, as well several clocks for meeting-houses.

Friend, Engell. New York, New York. 1825.

Friend, George. New York, New York. 1820.

Frost, Jonathan. Reading, Massachusetts. 1856.

Frost & Mumford. Providence, Rhode Island. 1810.

Frost, Oliver. Providence, Rhode Island. 1800.

G

Gaines, John. Portsmouth, New Hampshire. 1800.

Galbraith, Patrick. Philadelphia, Pennsylvania. 1795-1811.

Galpin, Moses. Bethlehem, Connecticut. Before 1821. Was not a maker of clocks, but a peddlar who often put his own name on clocks he purchased from manufacturers.

Galt, Peter. Baltimore, Maryland. 1804.

Galt, Samuel. Williamsburg, Virginia. 1751.

Gardiner, B. New York, New York. 1832.

Gardiner, John B. Ansonia, Connecticut. 1820.

Garrett, Benjamin. Goshen, New York. 1820.

Garrett, Philip. Philadelphia, Pennsylvania. 1819.

Garrett & Sons, P. Philadelphia, Pennsylvania. 1832.

Gates, Zaccheus. Charlestown, Massachusetts. 1831.

Gaw, William P. Philadelphia, Pennsylvania. 1819.

Gaylord, Homer. Norfolk, Connecticut. Until 1812. Made clocks on the family farm until 1812, at which time a flood destroyed the dam. He later moved to Homer, New York.

Geddes, Charles. Boston, Massachusetts. 1773.

Gelston, George S. New York, New York. 1832.

Gelston, Hugh. Baltimore, Maryland. 1832.

Gerding & Siemon. New York, New York. 1832.

Gerrish, Oliver. Portland, Maine. 1834.

Gibbons, Thomas. Philadelphia, Pennsylvania. -?-

Gilbert, Jordan & Smith. New York City, 1832.

Gilbert Manufacturing Co., The. Winsted, Connecticut. Incorporated 1866-1871. After reorganization in 1871, the business was called the Wm. L. Gilbert Clock Company. The company is no longer in business, and the factory building has been made into condominiums.

Gilbert, William L. Winsted, Connecticut. 1823-1866. Gilbert joined Lucius Clarke in 1841 or 1842, the firm name being Clarke, Gilbert & Co. It later became W. L. Gilbert, and in 1866 was incorporated as The Gilbert Manufacturing Co. Reorganized in 1871 as the Wm. L. Gilbert Clock Co.

Gill, Caleb. Hingham, Massachusetts. 1785.

Gill, Leavitt. Hingham, Massachusetts. 1785.

Giraud, Victor. New York, New York. 1847.

Glover, William. Boston, Massachusetts. 1823.

Goddard, George S. Boston, Massachusetts. 1823.

Godfrey, William. Philadelphia, Pennsylvania. 1750-1763.

Godschalk, Jacob. Philadelphia, Pennsylvania. -?-

Goff, Charles. -?-?-

Goodfellow, William. Philadelphia, Pennsylvania. 1796-1799.

Goodfellow, William. Philadelphia, Pennsylvania. 1813.

Goodfellow, William & Son. Philadelphia, Pennsylvania. 1796-1799.

Goodhue, D.T. Providence, Rhode Island. 1824.

Goodhue, Richard S. Portland, Maine. 1834.

Gooding, Henry. Boston, Massachusetts. 1810-1830.

Gooding, Joseph. Dighton, Massachusetts and Bristol, Rhode Island. -?-

Goodrich, Chauncey. Bristol, Connecticut. 1856.

Goodrich, G. Forestville, Connecticut. 1860.

Goodwin, Horace, Jr. Hartford, Connecticut. 1831-1841.

Gorden, Smyley. Lowell, Massachusetts. 1832. "He was a maker of clock cases, and put his name on them."

Gordon, Thomas. Boston, Massachusetts. 1759. "From London, opposite the Merchant's Coffee House, sells all kinds of Timepieces."

Gould, Abijah. Rochester, New York. 1834.

Govett, George. Philadelphia, Pennsylvania. 1813-1831.

Grant, James. Hartford & Wethersfield, Connecticut. A London-trained clockmaker, Grant settled in Hartford in 1794. Sales being slow, in 1795, he advertised as solely a mender and repairer of clocks; in 1796, he moved to Wethersfield where he continued repairing clocks.

Grant, William. Boston, Massachusetts. About 1815.

Graves, Alfred. Willow Grove, Pennsylvania. 1845.

Green, John. Carlisle, Pennsylvania. -?-

Green, John. Philadelphia, Pennsylvania. 1794.

Greenleaf, David. Hartford, Connecticut. 1799. Born in Norwich in 1765 and served his apprenticeship under Thomas Harland. Settled in Hartford in 1788 where he opened a shop, making clocks, silverware, repairing watches and selling jewelry. In 1796, he gave up making clocks, limiting himself to the repair of watches, and, it is presumed, the sale of clocks made by Harland. In 1811, he closed his shop and took up dentistry. He died in Hartford in 1835.

Greenough, N. C. Newburyport, Massachusetts. 1848.

Gridley, Timothy. Sanbornton, New Hampshire. 1808. He established a factory for the manufacture of wooden clocks, putting the business in the hands of two men from Connecticut, Mr. Peck and Mr. Holcomb. Mr. Gridley was later bought out by Col. Simon Cates. Painting and lettering of clock faces was done by a Mrs. James Connor.

Griffin, Henry. New York, New York. 1793-1818.

Griffith, Edward. Litchfield, Connecticut. Emigrated from England and settled in Litchfield in 1790. Later in the year, he moved to Savannah, Georgia, where he advertised as a watchmaker.

Griffith, Owen. Philadelphia, Pennsylvania. -?-

Griswold, Daniel White. East Hartford, Connecticut, b. 1767; d. 1844. In 1782, Griswold was apprenticed to his uncle, Timothy Cheney, clockmaker. In 1788, he had established his own business as a clockmaker in East Hartford, but presumably gave up the trade before 1800. He later acted as a trader between Boston and New York.

Groppengerser, J.L. Philadelphia, Pennsylvania. 1840.

Grotz, Isaac. Easton, Pennsylvania. 1810-1835.

Gruby, Edward L. Portland, Maine. 1834.

Guild, Jeremiah. Cincinnati, Ohio. 1831.

Guile, John. Philadelphia, Pennsylvania. 1819.

Guinard, F.E. Baltimore, Maryland. 1817.

H

Haas & Co., John. New York, New York. 1825. Makers of musical clocks.

Hall, John. Philadelphia, Pennsylvania. 1811-1819.

Hall, Seymour & Co. Unionville, Connecticut. About 1820.

Ham, George. Portsmouth, New Hampshire. 1810.

Ham, Supply. Portsmouth, New Hampshire. A member of the clockmaking fraternity of Portsmouth.

Hamlen, Nathaniel. Augusta, Maine. 1795-1820.

Hamlin, William. Providence, Rhode Island. 1797.

Hampton, Samuel. Chelsea, Massachusetts. 1847.

Hanks, Benjamin. Mansfield and Litchfield, Connecticut. 1778-1808. Born in Mansfield in 1755, Hanks was a natural mechanic, both skillful and energetic. He made clocks, was a goldsmith, a maker of stockings and looms, caster of bells and brass cannons, and a maker of compasses. In 1780, he moved to Litchfield, living at 82 South Street, where he continued in his various business ventures. He eventually returned to Mansfield, where he continued to make clocks and bells, and also carried on a woolen trade. He died in 1824 in Troy, New York.

Harden, James. Philadelphia, Pennsylvania. 1819.

Harland, Thomas. Norwich, Connecticut. 1773-1807. Born in England in 1753, he learned the clock and watchmaking trade, traveling extensively in England, as well as on the continent. Due to adverse conditions in the English trade, he emigrated to the colonies in 1773.

Arriving in Boston at the time of the Boston Tea Party, he changed his plans of settling in Boston town, going immediately to Norwich where he established a clock business. One of his earlier customers was Nathan Hale, the Connecticut patriot. Due to the local policy of not importing English goods, including clocks, Harland prospered; by 1790, he was employing ten or twelve apprentices from all parts of New England. According to Penrose Hoopes, "Contrary to tradition, Eli Terry was not a Harland apprentice."

His shop burned to the ground in 1795, and, though aged 60 at the time, Harland immediately re-established himself in a new location. After his death in 1807, his importance was fully realized; he was one of the most outstanding figures in the history of early American clockmaking. Harland was very well educated and was an extremely skillful mechanic, producing clocks and watches superior in workmanship to those made by his contemporaries. His greatest influence upon clockmaking in America was in the great number of apprentices that he trained and then sent forth to practice their trade.

Harrison, James. Waterbury, Connecticut. 1790-1830. Born in Litchfield, Connecticut in 1767, it is presumed he learned his trade from his uncle, Timothy Barnes, clockmaker. He practiced his trade in Southington, and, in 1795, moved to Waterbury. In 1800, he opened a small shop on North Main Street on the Little Brook, using a water wheel for power, said to be the first used in the town of Waterbury. A great mechanic, but a terrible business man, he soon lost his business. He eventually ended in New York City, where he died in poverty.

Harrison, John. Philadelphia, Pennsylvania. -?-

Harrison, Wooster (brother of James). Trumbull and Newfield, Connecticut. b. 1772; d.? Settled in Trumbull in or about 1795 and commenced the trade of clockmaking; in June of 1800, he announced his move to the town of Newfield.

Hart, Alpha. Goshen, Connecticut. 1820.

Hart, Eliphaz. Norwich on the Green, Connecticut. 1812.

Hart, Henry (brother of Alpha). Hart Hollow, Goshen, Connecticut. -?-

Hart, Judah. Norwich at the Landing, Connecticut. 1812.

Hart, Orrin. Bristol, Connecticut. 1840.

Hart & Wilcox. Norwich, Connecticut. -?-

Harwood, George. Rochester, New York. 1839.

Haselton & Wentworth. Lowell, Massachusetts. 1832.

Hatch, George D. North Attleborough, Massachusetts. 1856.

Hawxhurst (possibly Hauxhurst) & Demilt. New York, New York. 1790.

Hawxhurst, Nathaniel. New York, New York. 1786-1798.

Hayes, Peter B. Poughkeepsie, New York. 1831.

Heath, Reuben. Scottsville, New York. 1791-1818. Clockmaker and repairer of watches; also sold clocks of other makes.

Hedge, George. Buffalo, New York. 1831.

Heffords, ——. Middleboro, Massachusetts. Famous for the clocks he invented and produced, which were of superior quality

Heiligg, Jacob. Philadelphia, Pennsylvania. 1770-1824.

Heilig, John. Germantown, Pennsylvania. 1824-1830.

Hendrick, Barnes & Co. Forestville, Connecticut. 1845. Utilizing the old Ives Co. shop, this company made the first ever marine clocks.

Hendricks, Uriah. New York, New York. 1756. By trade, a watchmaker.

Hepton, Frederick. Philadelphia, Pennsylvania. 1785.

Hequembourg, C. New Haven, Connecticut. 1818. In business in New Haven for an undetermined number of years, where he sold and repaired clocks and watches.

Heron, Isaac. New York, New York. 1769-1780.

Herr, William, Jr. Providence, Rhode Island. 1849.

Hicks, Willet. New York, New York. 1790.

Hildeburn, Samuel. Philadelphia, Pennsylvania. 1819.

Hildeburn & Watson. Philadelphia, Pennsylvania. 1832. "Manufacturers of Jewelry and Watch Case Makers, and Importers of Watches and Fancy Goods."

Hildeburn & Woodworth. Philadelphia, Pennsylvania. 1819.

Hildreth, Jonas. Salisbury, Vermont. 1805.

Hill, D. Reading, Pennsylvania. 1820-1840.

Hill, Joakim. Flemington, New Jersey. 1800.

Hiller, Joseph. Salem, Massachusetts. 1770. Opened a shop on the Exchange opposite the Court House, after moving from Boston.

Hilldrop (or Hilldrup), Thomas. Hartford, Connecticut. 1774-1794. Emigrated from London and set up shop as a watchmaker, jeweler and silversmith.

Hills, Amariah. New York, New York. 1845.

Hills, George. Plainville, Connecticut. 1842.

Hills, Goodrich & Co. Plainville, Connecticut. 1841-1846.

Hitchcock, H. Lodi, New York. About 1800.

Hoadley, Samuel & Luther. Winsted, Connecticut. 1807-1813. In 1807, the brothers Hoadley, in company with Riley Whiting, opened a plant in Winsted for making of wooden clocks. Luther died in 1813, while Samuel left the business and entered the army.

Hoadley, Silas. Plymouth, Connecticut. 1808-1849. Taught the carpenter's trade by his uncle, Calvin Hoadley, Silas went to work for Eli Terry, and, in 1809, entered into partnership with Terry and Seth Thomas at Greystone. Terry left the business in 1810, with Thomas leaving in 1812. Hoadley kept the business going until his retirement in 1849.

Hodges, Erastus. Harwinton, Connecticut. 1831-1842.

Hodges & North. Wolcotville (now Torrington), Connecticut. 1830.

Hodgson, William. Philadelphia, Pennsylvania. 1785.

Hoffner, Henry. Philadelphia, Pennsylvania. 1791.

Holbrook, George. Brookfield, Massachusetts. 1803. Made the clock and bell used in the Leicester, Massachusetts meeting-house, 1803.

Holbrook,———. Medway, Massachusetts. About 1830.

Hollinshead, Jacob. Salem, Massachusetts. 1771.

Hollinshead, Morgan. Moorestown, New Jersey. -?-

Holman, Salem. Hartford, Connecticut. 1816.

Holway, Philip. Falmouth, Massachusetts. 1800.

Homer, William. Moreland, Pennsylvania. 1849.

Hood, Francis. New York, New York. 1810.

Hood, John. Philadelphia, Pennsylvania. -?-

Hooker & Goodenough. Bristol, Connecticut. 1849.

Hopkins & Alfred. Harwinton (1820) and Hartford (1827), Connecticut. Makers of outstanding wooden clocks.

Hopkins, Asa. Litchfield, Connecticut. 1820 and earlier. Received a patent in 1813 on an engine designed for cutting wheels.

Hopkins, Henry P. Philadelphia, Pennsylvania. 1832.

Horn, Eliphalet. Lowell, Massachusetts. 1832.

Horn, E.B. Boston, Massachusetts. 1847.

Hotchkiss & Benedict. Auburn, New York. About 1820. Makers of shelf clocks.

Hotchkiss, Elisha. Burlington, Connecticut. About 1815.

Hotchkiss & Field. Burlington, Connecticut. 1820.

Hotchkiss, Hezekiah. New Haven, Connecticut. 1748. A Yankee jack-of-all-trades, Hotchkiss learned the clockmaking trade at an early age, opening his own shop in New Haven at the age of nineteen. He died early at the age of thirty two from an inoculation for smallpox.

Hotchkiss, L.S. New York. 1840.

Hotchkiss & Pierpont. Plymouth, Connecticut. 1811-?

Hotchkiss, Robert & Henry. Plymouth, Connecticut. Before 1846.

Hotchkiss, Spencer & Co. Salem Bridge (now Naugatuck), Connecticut. 1832. Manufacturers of brass eight-day clocks.

Howard & Davis. Boston, Massachusetts. 1847-1856.

Howard, Edward. Roxbury, Massachusetts. 1840-1882. Born in Hingham, Massachusetts in 1813. Started business for himself in 1840, taking David P. Davis on as a partner in 1847. For years they carried on a successful business making clocks and regulators. They were joined by Aaron L. Dennison in 1849, at which time they started a watch factory in Roxbury. The business moved to Waltham in 1854, and clocks made at this location were marked "Dennison, Howard and Davis." In 1857, the Waltham factory was sold to Royal E. Robbins. Howard returned to Roxbury and opened the old factory of the Boston Watch Co., eventually putting his product on the market as "Howard Watches," which established an excellent reputation as timekeepers.

Howard, Thomas. Philadelphia, Pennsylvania. 1789-1791.

Howe, Jubal. Boston, Massachusetts. 1833.

Howell, Nathan. New Haven, Connecticut, b. 1740; d. 1784. Maker of brass clocks.

Hoyt, George A. Albany, New York. 1830.

Hoyt, James A. Troy, New York. 1837.

Hubbard, Daniel. Medfield, Massachusetts. 1820.

Hubbell, L. -?-?-

Huckel, Samuel. Philadelphia, Pennsylvania. 1819.

Huguenail, Charles T. Philadelphia, Pennsylvania. 1799.

Humbert, Dross. Philadelphia, Pennsylvania. 1795.

Hunt, ——. New York, New York. 1789.

Hurtin & Burgi. Bound Brook, New Jersey. 1766.

Hutchins, Abel. Concord, New Hampshire. 1788-1819. Apprenticed to Simon Willard, eventually ending up in business with his brother Levi.

Hutchins, Levi. Concord, New Hampshire. 1786-1819. Like his brother Abel, Levi was also an apprentice of Simon Willard. He learned his trade of repairing watches in Abington, Connecticut. In 1786, he started his business of making brass clocks, and, in 1788, was joined by his brother Abel.

Hyman, Samuel. Philadelphia, Pennsylvania. 1799.

I

Ingersoll, Daniel G. Boston, Massachusetts. 1800-1810.

Ingraham, Elias. Bristol, Connecticut. 1835-1885. Designer of many styles of clock cases, including the "Sharp Gothic" style. Originally a cabinet maker, Ingraham made clock cases for George Mitchell, at the same time learning the clockmaking trade at Mitchell's factory. In 1835, he struck out on his own, buying a factory in Bristol and starting the manufacture of clocks; he later formed a partnership with his brother and Elisha C. Brewster. This business, Brewster & Ingraham, was succeeded by E. & A. Ingraham, which was also succeeded by E. Ingraham & Co. in 1856. The company went public in 1881, with the business being carried on from that time.

Ingraham, E. & A. Bristol, Connecticut. 1848-1855.

Ingraham & Co., E. Bristol, Connecticut. 1856-1881. Joint stock company formed in 1881 and called The E. Ingraham Co.

Ives & Birge. Bristol, Connecticut. About 1843.

Ives Brothers. Bristol, Connecticut. 1815-1820, 1822-1837.

Ives, Charles G. Bristol, Connecticut. 1810-1820. Maker of wooden clocks.

Ives, Chauncey. Bristol, Connecticut.

Ives, C. & L.C. Bristol, Connecticut. 1832. In approximately 1830, a factory was built in Bristol by Chauncey and Lawson C. Ives to make eight-day brass clocks, based on a design invented by John Ives. After a number of successful years in business, the factory closed in 1836.

Ives, Ira. Bristol, Connecticut. about 1815-1837.

Ives, Joseph. Bristol, Connecticut. and New York, New York. 1811-1825. Ives started as a maker of wooden clocks; in 1818, he invented a rather large, clumsy, and mostly unsuccessful brass and iron clock, only a few of which were made.

Ives Lawson.C. Bristol, Connecticut. 1827-1836.

J

Jacks, James. Philadelphia, Pennsylvania. -?-

Jackson, Joseph H. Philadelphia, Pennsylvania. 1802-1810.

James, Joshua. Boston, Massachusetts. 1823.

Jeffreys, Samuel. Philadelphia, Pennsylvania. -?-

Jencks, John E. Providence, Rhode Island. 1800.

Jenkins, Harman. Albany, New York. 1813.

Jerome, Chauncey. Bristol and New Haven, Connecticut. 1816-1860. Jerome started working for Eli Terry in the winter of 1816 and a little later started making clocks by himself. This led to a move to Bristol in 1821, and, in 1824, the firm of "Jeromes and Noble" was formed, consisting of Chauncey and Noble Jerome and Elijah Darrow. Shortly after the start of the company, Chauncey Jerome came up with his "Bronze Looking Glass Clock," with which the firm did very well until the panic of 1837. It was in 1838, with the invention of the brass one-day clock, that Jerome drove wooden clockmakers out of business.

Business grew, and the firm made many shipments to England, where the superiority of his product was appreciated. The plant was moved to New Haven in 1844, and, in 1855 the company went public: "The Jerome Manufacturing Company." It was at this time that P. T. Barnum was associated with the business, during the last six months of its existence. What drove the business under was the assumption by the Jerome

Manufacturing Company of the indebtedness of the Terry & Barnum Co. Jerome spent his last years in penniless obscurity.

Jerome, Chauncey & Noble. Richmond, Virginia and Hamburg, South Carolina. 1835-1836. Cases and parts were made in Bristol, Connecticut, then shipped to both locations where they were assembled by Connecticut workmen sent there for that purpose.

Jerome & Darrow. Bristol, Connecticut. 1824-1831. "Manufacturers of Thirty Hour and Eight Day Wood Clocks."

Jerome & Grant. Bristol, Connecticut. -?-

Jerome Manufacturing Company New Haven, Connecticut. 1850-1855.

Jerome, Noble. Bristol, Connecticut. 1820-1840. Brother and partner of Chauncey Jerome.

Jewell, Jerome & Co. Bristol, Connecticut. 1849. Makers of town clocks.

Job, John. Philadelphia, Pennsylvania. 1819.

Jocelin, Simeon. New Haven, Connecticut. b. 1746; d.1823. Most likely Jocelin was apprenticed to Isaac Doolittle to learn the clockmaking trade. He first appeared in business in 1768, quietly pursuing the trade until 1776, at which point the Revolution made the demand for clocks very limited. Jocelin followed other pursuits until 1790, at which time he re-opened his clock trade.

In March of 1800, Jocelin received a patent for a "Silent Moving Time Piece," being produced both in tall case and shelf size. With this, business prospered, and Jocelin continued active in the business up to the time of his death.

Jocelyn, Nathaniel. New Haven, Connecticut. 1790.

Johnson, Addison. Wolcottville (now Torrington), Connecticut. 1825. Manufacturer of pillar and scroll clocks, with quality works and handsome cases.

Johnson, Chauncey. Albany, New York. 1829. Known for "Musical, ornamental and common clocks."

Johnson, Simon. Sanbornton, New Hampshire. 1830-1860. "A very superior quality of clock has been produced from this establishment by the senior Mr. Johnson and latterly by the Johnson Brothers."

Johnson, William S. New York, New York. About 1830.

Jonckheere, Francis. Baltimore, Maryland. 1817.

Jones, Abner. Weare, New Hampshire. 1780. Made old fashioned eight-day brass clocks.

Jones, Ball and Poor. Boston, Massachusetts. 1847.

Jones, Edward K. Bristol, Connecticut. 1825.

Jones, Ezekiel. Boston, Massachusetts. 1823.

Jones, George. Wilmington, Delaware. 1814.

Jones, George Jr. Wilmington, Delaware. 1814.

Jones, Jacob. Baltimore, Maryland. 1817.

Jones, Jacob. Concord, New Hampshire. 1820.

Jones, Samuel. Baltimore, Maryland. 1817.

Jones & Woods. 1850-1860.

Joseph, Isaac. Boston, Massachusetts. 1823.

Joslyn, James. New Haven, Connecticut. 1798-1820.

K

Kearney, Hugh. Wolcotville, Connecticut. 1830.

Kedzie, J. Rochester, New York. 1847.

Kellogg, Daniel. Hebron, Connecticut, b. 1766; d. 1855. After serving an apprenticeship under Daniel Burnap, Kellogg opened his own shop in 1787 in the town of Hebron. His clockmaking activities apparently ceased after 1800.

Kelly (or Kelley), Allen. Sandwich, Massachusetts. 1810-1830.

Kelly, Ezra. New Bedford, Massachusetts. 1823-1845.

Kelly, John. New Bedford, Massachusetts. 1836.

Kemble, William. New York, New York. 1786.

Kemlo, Francis. Chelsea, Massachusetts. 1847.

Kennard, John. Newfields, New Hampshire. 19th century. Made brass clocks, the clock cases being made by Henry Wiggins, Jr.

Kennedy, Elisha. Middletown, Connecticut, b. 1766; d.?

Kennedy, Patrick. Philadelphia, Pennsylvania. 1795-1799.

Kennedy, T. Connecticut. 1860.

Kenney, Asa. West Millbury, Massachusetts. About 1800.

Kepplinger, Samuel. Baltimore, Maryland. 1800.

Kerner & Paff. New York, New York. 1796. Engaged in the sale of cuckoo clocks and musical figural clocks.

Ketcham & Hitchcock. New York, New York. 1818.

Kimball, John Jr. Boston, Massachusetts. 1823.

Kimberly, Roswell. Ansonia, Connecticut. 1850.

Kincaird, Thomas. Christiana Bridge, Delaware. 1775.

Kippen, George. Bridgeport, Connecticut. 1822. In addition to making and repairing clocks and watches, Kippen also merchandised a large variety of commodities.

Kirk, Charles. Bristol, Connecticut. 1823-1833. After manufacturing clocks for approximately ten years, Kirk sold his factory to Elisha C. Brewster. He remained on for a further five years as Brewster's superintendent.

Kirk, Charles. New Haven, Connecticut. 1847. Manufacturer of brass marine clocks.

Kline, B. Philadelphia, Pennsylvania. 1841.

Kline, John. Philadelphia, Pennsylvania. 1820.

Kline, John. Reading, Pennsylvania. 1820-1840.

Knowles, John. Philadelphia, Pennsylvania. -?-

Kohl, Nicholas. Willow Grove, Pennsylvania. 1830.

Koplin, Washington. Norristown, Pennsylvania. 1850.

Kroeber, F. New York. 1890.

Kumbell, William. New York, New York. 1775-1789.

L

Labhart, W.I. New York, New York. 1810.

Ladomus, Lewis. Philadelphia, Pennsylvania. 1846.

Lamb, Cyrus. Oxford, Massachusetts. 1832. Known as a millwright and skilled mechanic, it is presumed that Lamb was not a full-time clockmaker. Legend has it that at the time of a fire in his shop in 1832, he had a most remarkable clock of his own design which would presumably run for a number of years with only one winding.

Lamione, A. Philadelphia, Pennsylvania. 1811.

Lamson, Charles. Salem, Massachusetts. 1850. (in conjunction with James Balch).

Lane, James. Philadelphia, Pennsylvania. 1813.

Lane, J. Southington, Connecticut. -?-

Lane, Mark. Southington, Connecticut. 1831. "Manufacturer of Eli Terry's Patent Clocks."

Langdon, Edward. Bristol, Connecticut. 19th century. Associated for a time with S. Emerson Root.

Lanny, D.F. Boston, Massachusetts. 1789.

Laquine, ———. Philadelphia, Pennsylvania. -?-

Larkin, Joseph. Boston, Massachusetts. 1841-1847.

Latimer, James. Philadelphia, Pennsylvania. 1819.

Laundry, Alexander. Philadelphia, Pennsylvania. -?

Launey, David. New York, New York. 1801.

Lawrence, George. Lowell, Massachusetts. 1832.

Lawson, William H. Waterbury, Connecticut. -?-

Leach, Caleb. Plymouth, Massachusetts. 1776-1790.

Leach & Bradley. Utica, New York. 1832.

Leavenworth, Mark. Waterbury, Connecticut. 1810-1830.

Leavenworth & Co., Mark. Waterbury, Connecticut. 1832.

Leavenworth & Sons. Albany, New York. 1817.

Leavenworth, William. Waterbury, Connecticut. 1802-1815. In 1802, he had a shop on the Mad River for the making of clocks.

Leavitt, Dr. Josiah. Hingham, Massachusetts. 1772.

Lee, William. Charlestown, South Carolina. 1717.

Lefferts, Charles. Philadelphia, Pennsylvania. 1819.

Lefferts & Hall. Philadelphia, Pennsylvania. 1819.

Le Huray, Nicholas, Jr. Philadelphia, Pennsylvania. 1832.

Leigh, David. Pottstown, Pennsylvania. 1849.

Lemist, William King. Dorchester, Massachusetts. 1812. An apprentice of Simon Willard, Lemist was lost at sea in 1820.

Lescoiet, Lanbier. Hartford, Connecticut. An itinerant watch and clockmaker, Lescoiet established a shop in Hartford in 1769 and lasted in business approximately two years.

Leslie & Price. Philadelphia, Pennsylvania. 1793-1799.

Leslie, Robert. Philadelphia, Pennsylvania. 1745-1791.

Lester, Robert. Philadelphia, Pennsylvania. 1791-1798.

Le Tilier, John. Philadelphia, Pennsylvania. -?-

Levi, Isaac. Philadelphia, Pennsylvania. -?-

Levy, Michael. Philadelphia, Pennsylvania. 1813.

Lewis, Erastus. New Britain (later Waterbury), Connecticut. About 1800.

Lewis, Levi. Bristol, Connecticut. 1811-1820.

Liebert, Henry. Norristown, Pennsylvania. 1849.

Limeburner, John. Philadelphia, Pennsylvania. 1791.

Lind, John (or Johannes). Philadelphia, Pennsylvania. 1791.

Lister, Thomas. Halifax, British North America. 1760-1802. Produced a line of exquisitely made tall clocks.

Litchfield Manufacturing Co. Litchfield, Connecticut. 1850.

Little, Peter. Baltimore, Maryland. 1799.

Little & Eastman. Boston, Massachusetts. 1900.

Lockwood & Scribner. New York, New York. 1847.

Lohse & Keyser. Philadelphia, Pennsylvania. 1832. Clock importers.

Lord & Goddard. Rutland, Vermont. 1797-1830.

Lorse, Miles. Plymouth Hollow, Connecticut. c. 1855.

Lorton, William B. New York, New York. 1810-1825. "Manufacturer and wholesale dealer in American clocks in all their variety."

Lovell Manufacturing Co. Erie, Pennsylvania. 1880.

Lovis, Capt. Joseph. Hingham, Massachusetts. 1775-1804.

Low & Co., John J. Boston, Massachusetts. 1832.

Lowens, David. Philadelphia, Pennsylvania. 1785.

Lowrey, David. Newington, Connecticut, b. 1740; d. 1819. It is possible that this skilled craftsman served an apprenticeship under Ebenezer Balch. He set up shop in Newington, and remained there for the rest of his life, producing quality clocks and doing the blacksmithing work that the community required.

Ludwig, John. Philadelphia, Pennsylvania. 1791.

Lufkin & Johnson. Boston, Massachusetts. 1800-1810.

Lukens, Isaiah. Philadelphia, Pennsylvania. 1790-1828.

Lukens, Seneca. Horsham Meeting, Pennsylvania. 1830.

Luscomb, Samuel. Salem, Massachusetts. 1773. Made the clock installed in the East Meeting House in 1773.

Lux Clock Manufacturing Co. Waterbury, Connecticut. 1930.

Lyman, G.E. Providence, Rhode Island. 1849.

Lyman, Roland. Lowell, Massachusetts. 1832.

Lynch, John. Baltimore, Maryland. 1804-1832. "Manufacturer of Silver Work and Clock and Watch Maker."

M

Macfarlane, John. Boston, Massachusetts. 1800-1810.

Mackay, Crafts. Boston, Massachusetts. 1789.

Manning, Richard. Ipswich, Massachusetts. 1748-1760.

Manross, Elisha. Bristol (village of Forestville), Connecticut. 1827-1849. Manross set up shop in the John Ives plant in Forestville; in 1845 he built his own factory close by the railroad.

Manross (two brothers). Bristol, Connecticut. 1860. Involved in the manufacture of marine clock movements.

Marache, Solomon. New York, New York. 1759.

Marand, Joseph. Baltimore, Maryland. 1804.

Marble, Simeon. New Haven, Connecticut. 1817. Maker and retailer of clocks, watches, and silverware.

Marine Clock Manufacturing Co. New Haven, Connecticut. 1847.

Marks, Isaac. Philadelphia, Pennsylvania. 1795.

Marquand & Bros. New York, New York. 1832.

Marsh, George. Bristol, Connecticut. -?-

Marsh, George C. Wolcottville (now Torrington), Connecticut. 1830.

Marsh, Gilbert & Co. Farmington, Connecticut. 1820. Makers of shelf clocks.

Marshall & Adams. Seneca Falls, New York. c. 1825.

Masi & Co. Washington, D.C. 1833.

Masi, Seraphim. Washington, D.C. 1832.

Mason, H.G. Boston, Massachusetts. 1844-1849.

Mathey, Lewis. Philadelphia, Pennsylvania. 1797.

Matlack, White. Philadelphia, Pennsylvania. -?-

Matlack, White C. New York, New York. 1769-1775.

Matlack, William. Philadelphia, Pennsylvania. -?-

Matthews & Jewel. Bristol, Connecticut. c. 1850.

Matthewson, J. Providence, Rhode Island. 1849.

Maurepas,———. Bristol, Connecticut. 1855.

Maus, Frederick. Philadelphia, Pennsylvania. 1785-1793.

Mayer, Elias. Philadelphia, Pennsylvania. 1832.

Maynard, George. New York, New York. 1702-1730.

McClure, John. Boston, Massachusetts. 1823

M'Cormick, Robert. Philadelphia, Pennsylvania. -?-

McDowell, James. Philadelphia, Pennsylvania. 1794-1799 or perhaps later.

McDowell, James, Jr. Philadelphia, Pennsylvania. 1805-1825.

M'Harg, Alexander. Albany, New York. 1817.

McIlhenny, Joseph E. Philadelphia, Pennsylvania. 1819.

McIlhenny & West. Philadelphia, Pennsylvania. 1819.

M'Keen, H. Philadelphia, Pennsylvania. 1832.

McMyers, John. Baltimore, Maryland. 1799.

Mead, Adriance & Co. Ithaca, New York. 1832.

Mead, Benjamin. Castine, Maine. 1800-1810.

Meeks, Edward, Jr. New York, New York. 1796. "Makes and has for sale 8 day clocks and chiming timepieces."

Melcher,———. Plymouth Hollow (now Thomaston), Connecticut. About 1790.

Melly, "Brothers Melly." New York, New York. 1829. "All kinds of clocks and watches."

Mendenhall, Thomas. Philadelphia, Pennsylvania. -?-

Mends, Benjamin. Philadelphia, Pennsylvania. -?-

Mends, James. Philadelphia, Pennsylvania. 1795.

Mezies, James. Philadelphia, Pennsylvania. 1800 and later.

Merchant, William. Philadelphia, Pennsylvania. -?-

Merriam, Silas. Bristol, Connecticut. 1790.

Merriman & Bradley. New Haven, Connecticut. 1825.

Merriman, Silas. New Haven, Connecticut, b. 1734; d. 1805. Born in Wallingford in 1734, Merriman was probably apprenticed to Macock Ward. In 1760, he established his home and his shop on State Street, New Haven, where he pursued his trade as a maker of brass clocks and silversmith.

Merriman, Titus. Bristol, Connecticut. 1810-1830. Connected with Butler Dunbar in the making of clocks in 1810.

Mery (or Merry), F. Philadelphia, Pennsylvania. 1799.

Meyer, J.A. New York, New York. 1832.

Meyers, John. Fredericktown, Maryland. 1793-1825.

Millard, Squire. Warwick, Rhode Island. Maker of hall clocks.

Miller, Aaron. Elizabethtown, North Carolina. (?) 1747.

Miller, Abraham. Easton, Pennsylvania. 1810-1830.

Miller, Edward F. Providence, Rhode Island. 1824-1849.

Miller (or Millar), Thomas. Philadelphia, Pennsylvania. 1832.

Milne, Robert. M. New York, New York. 1798-1802.

Minor, Richardson. Danbury, Connecticut, b. 1763; d. 1797. Born and raised in Stratford, he commenced work as a clockmaker and goldsmith in 1758, continuing until his death in 1797.

Mitchell & Atkins. Bristol, Connecticut. 1830.

Mitchell, George. Bristol, Connecticut. 1827-1840.

Mitchell, Hery. New York, New York. 1786-1802.

Mitchell, Hinman & Co. Bristol, Connecticut. 1831.

Mitchell & Mott. New York, New York. 1793-1802.

Mitchell, Phineas. Boston, Massachusetts. 1823.

Mitchelson, David. Boston, Massachusetts. 1774.

Monroe & Co., E. & C.H. Bristol, Connecticut. 1856.

Monroe, John. Barnstable, Massachusetts. -?-

Montgomery, Robert. New York, New York. 1786.

Moolinger, Henry. Philadelphia, Pennsylvania. 1794.

Morgan, Elijah. Poughkeepsie, New York. 1832.

Morgan, Theodore. Salem, Massachusetts. 1837.

Morgan, Thomas. Baltimore, Maryland. 1774.

Morgan, Thomas. Philadelphia, Pennsylvania. 1779-1793.

Morrell, Benjamin. Boscawen, New Hampshire. 1816-1845.

Morrell & Mitchell. New York, New York. 1816-1820.

Morris, William. Grafton, Massachusetts. 1765-1775.

Morse & Co.(Myles Morse & Jeremiah Blakeslee). Plymouth Hollow (now Thomaston), Connecticut. 1841-1849. Makers of clocks with brass works, 1 day time and wire gong.

Morse, Elijah (son-in-law of William Crane). Canton, Massachusetts. 1819-?

Morse, Henry (son-in-law of William Crane). Canton, Massachusetts. 1819.

Morse, Miles (or Myles). Plymouth, Connecticut. 1849. Associated with Jeremiah Blakeslee in Morse & Co. prior to 1849. In partnership with Gen. Thomas A. Davis of New York City, he built a plant for producing clocks on the West Branch of the Naugatuck River, Plymouth, Connecticut, which was productive from 1850 to 1855.

Mosely, Robert E. Newburyport, Massachusetts. 1848.

Mott, Jordan. New York, New York. 1802-1825.

Moulton, Edward S. Rochester, New Hampshire. 1807.

Moulton, Francis. Lowell, Massachusetts. 1832.

Mulliken, Jonathan (son of Samuel) Newburyport, Massachusetts. 1774-1782.

Muller, N. New York. Case maker. 1875.

Mulliken, Joseph. Newburyport, Massachusetts, b. 1767; d. 1804.

Mulliken, Nathaniel. Lexington, Massachusetts. 1751-1789. His shop was burned by the British.

Mulliken, Samuel. Newburyport, Massachusetts. 1750-1756.

Mulliken, Samuel, Jr. Newburyport, Salem and Lynn, Massachusetts. 1781-1807. Presumed the grandson of Samuel, son of John and the nephew of Jonathan. Born in 1761, he was apprenticed to Jonathan Mulliken; he later opened a shop on State Street. On or about 1785, he moved to Salem, and then at a later date, to Lynn.

Munger, A. Auburn, New York. 1825. Noted for his shelf clocks incorporating pillars and looking glasses.

Munger & Benedict. Auburn, New York. 1833.

Munger & Pratt. Ithaca, New York. 1832.

Munroe, Daniel. Concord and Boston, Massachusetts. 1800-? Associated in business with his brother Nathaniel in the town of Concord.

Munroe, Daniel, Jr. Boston, Massachusetts. 1823.

Munroe, Nathaniel (brother of Daniel), Concord, Massachusetts. 1800-1817, Baltimore, Maryland. 1817-? Nathaniel served his apprenticeship with Abel Hutchins of Concord. In 1800, he went into business with his elder brother, Daniel, in Concord; Daniel moved to Boston in 1808, but Nathaniel stayed in Concord until 1817, at which time he moved to Baltimore, Maryland. He did quite a business, mainly in eight-day brass clocks, employing seven or eight apprentices and journeymen.

Munroe & Whiting (Nathaniel Munroe and Samuel Whiting), Concord, Massachusetts. 1808-1817.

N

Neal, Elisha. New Hartford, Connecticut. 1830.

Neiser (or Neisser), Augustine. Georgia and Germantown, Pennsylvania. 1736-1780. In 1736, he emigrated from Moravia to the state of Georgia; in 1739, he moved to Germantown, Pennsylvania. Apparently all of his clocks bear his name, but it is generally believed that none are dated.

Nettleton, Heath & Co. Scottsville, New York. 1800-1818.

Nettleton, W.K. Rochester, New York. 1834.

Newberry, James. Philadelphia, Pennsylvania. 1819.

Newell, Thomas. Sheffield, Massachusetts. 1810-1820.

New England Clock Co. Bristol, Connecticut. 1850.

New Haven Clock Co. New Haven, Connecticut. 1855. Succeeded the Jerome Manufacturing Co.

Nicholls, George. New York, New York. 1728-1750.

Nichols, Walter. Newport, Rhode Island. 1849.

Nicolet, Joseph Marci. Philadelphia, Pennsylvania. 1797.

Nicolette, Mary. Philadelphia, Pennsylvania. 1793-1799.

Ninde, James. Baltimore, Maryland. 1799.

Noble, Philander. Pittsfield, Massachusetts. -?-

Nolan & Curtis. Philadelphia, Pennsylvania. Painters of clock dials.

Nolen, Spencer. Philadelphia, Pennsylvania. 1819. Manufacturer of clock dials.

North, Norris. Wolcottville (now Torrington), Connecticut. 1820.

Northrop, R.E. New Haven, Connecticut. About 1820.

Northrop & Smith. Goshen, Connecticut. About 1820.

Norton, Samuel. Hingham, Massachusetts. 1785.

Norton, Thomas. Philadelphia, Pennsylvania. 1811.

Noyes, Leonard W. Nashua, New Hampshire. 1830-1840.

Nutter, E.H. Dover, New Hampshire. -?-

O

Oakes, Frederick. Hartford, Connecticut. 1828. Watchmaker and goldsmith.

Oakes, Henry. Hartford, Connecticut. 1839. The company dealt in watches, jewelry, cutlery, combs and fancy goods.

O'Hara, Charles. Philadelphia, Pennsylvania. 1799.

Oliver, Griffith. Philadelphia, Pennsylvania. 1785-1793.

Oliver, Welden. Bristol, Connecticut. About 1820. Maker of shelf clocks, wood works, one-day time, bell strike.

Olmstead, Nathaniel. New Haven, Connecticut. 1826. Shop situated at 117 North Side Chapel Street, a few doors east of the bank. Made and repaired clocks and watches.

One Hand Clock Co. Warren, Pennsylvania. 1890.

O'Neil, Charles. New Haven, Connecticut. 1823.

Orr, Thomas. Philadelphia, Pennsylvania. 1811.

Orten, Preston & Co. Farmington, Connecticut. About 1815.

Osgood, John. Boston, Massachusetts. 1823.

Osgood, John. Haverhill, New Hampshire. An early jeweler, Osgood also manufactured old-fashioned high clocks.

Owen, George B. New York. 1875.

Owen, Griffith. Philadelphia, Pennsylvania. 1813.

Oyster, Daniel. Reading, Pennsylvania. 1820-1840.

P

Packard, Isaac. North Bridgewater (now Brockton), Massachusetts. 18—. Was associated with Rodney Brace.

Packard, J. Rochester, New York. 1819.

Packard & Scofield. Rochester, New York. 1818. "Perpetual Motion." "Packard and Scofield, Watch-makers, have at their shop, next door south of 'The Telegraph,' a handsome assortment of Gold, Silver and Plated ware which will be sold at a moderate profit, for no man can live by the loss. Clocks and watches of every description repaired and warrented to keep in motion merely by winding every day."

Paine & Heroy. Albany, New York. 1813.

Palmer, John. Philadelphia, Pennsylvania. 1795.

Park, Seth. Park Town, Pennsylvania. 1790.

Parke, Solomon. Philadelphia, Pennsylvania. 1791-1819.

Parke & Co., Solomon. Philadelphia, Pennsylvania. 1799.

Parker, Gardner. Westborough, Massachusetts. 17—. A maker of clocks prior to the Revolutionary War.

Parker, Isaac. Deerfield, Massachusetts. 1780.

Parker, Thomas. Philadelphia, Pennsylvania. 1785-1813.

Parker & Co., Thomas. Philadelphia, Pennsylvania. 1819.

Parker, Thomas, Jr. Philadelphia, Pennsylvania. 1819.

Parmele, Abel. New Haven and Branford, Connecticut. Born in Guilford in 1703, he settled in New Haven in 1729. He apparently moved to Branford a few years later. Clocks by Parmele are presumed to still be in existence.

Parmele (or Parmelee, or Parmilee), Ebenezer. Guilford, Connecticut. 1690-1777. Credited with being the first clockmaker in the colony of Connecticut, he also carried on a fair trade in carpentry and cabinet-making. Made the clock for the church in Guilford, which was unquestionably the first steeple clock in Connecticut.

Parmier, John Peter. Philadelphia, Pennsylvania. 1793.

Parry, John J. Philadelphia, Pennsylvania. 1795-1813.

Patton, Abraham. Philadelphia, Pennsylvania. 1799.

Patton, David. Philadelphia, Pennsylvania. 1799.

Patton & Jones. Baltimore, Maryland. 1798.

Payne, Lawrence. New York, New York. 1732-1755.

Pearsall & Embree. New York, New York. 1786.

Pearsall, Joseph. New York, New York. 1786-1798.

Pearson, William, Jr. New York, New York. 1775.

Pease, Isaac T. Enfield, Connecticut. 1818.

Peck, Benjamin. Providence, Rhode Island. 1824.

Peck, Edson C. Derby, Connecticut. 1827.

Peck, Elijah. Boston, Massachusetts. 1789.

Peck & Co., Julius. Litchfield, Connecticut. 1820.

Peck, Moses. Boston, Massachusetts. 1789.

Peck, Timothy. Litchfield, Connecticut. 1790. Related by marriage to Miles Beach, it possible that Peck was also apprenticed to him. In 1787, Peck opened shop in Middletown, Connecticut, but returned to Litchfield in 1791, where he made and repaired clocks and watches.

Peckham & Knower. Albany, New York. 1814.

Perkins, Robinson. Jaffrey, N.H. -?-

Perkins, Thomas. Philadelphia, Pennsylvania. 1785-1799.

Perry, Marvin. New York, New York. 1769-1780. In 1776, Perry advertised as follows: "Repeating and Plain Clock and Watchmaker from London, where he has improved himself under the most eminent and capital artists in these branches, has opened shop in Hanover Square at the sign of the Dial. He mends and repairs, musical, repeating, quarterly, chiming, silent, pull, and common weight clocks."

Perry, Thomas. New York City 1749-1775. Advertised as follows: "Thomas Perry, watch-maker, from London, at the sign of the Dial, in Hanover Square, makes and cleans all sorts of clocks and watches in the best manner, and at a most reasonable rate."

Peters, James. Philadelphia, Pennsylvania. 1832.

Phillips, Joseph. New York, New York. 1713-1735.

Pierret, Matthew. Philadelphia, Pennsylvania. 1795.

Pierson, Henry S. Portland, Maine. 1834.

Pinkard, Jonathan. Philadelphia, Pennsylvania. -?-

Pitkin, Levi. East Hartford, Connecticut. 1795. In 1774, Levi Pitkin was born into the famous family of well-known Connecticut manufacturers. A clever mechanic, he was a jeweler, silversmith and clockmaker; talk has it that there are still a number of his clocks running in the East Hartford area. Pitkin Street in East Hartford was named after the family.

Pitkins (four brothers), East Hartford, Connecticut. 1826-1841 or later. In 1826, utilizing the same building, John O. Pitkin and Walter Pitkin started making silverware, while brothers Henry Pitkin and James F. Pitkin commenced manufacturing watches. These were the first watches made in America. All of the products were sold through the family business in Hartford.

Pitman & Dorrance. Providence, Rhode Island. 1800.

Pitman, Saunders. Providence, Rhode Island. 1780.

Pitman, William R. New Bedford, Massachusetts. "Manufacturers of gold and silverware, clocks and watches repaired and warrented."

Platt, ——. New Milford, Connecticut. 1793.

Platt, A.S. Bristol, Connecticut. 1849-1856.

Platt & Blood. Bristol, Connecticut. 1849.

Platt, G.W. & N.C. New York, New York. 1832.

Pomeroy, Chauncy. Bristol, Connecticut. About 1835.

Pomeroy, Noah. Bristol, Connecticut. 1849-1878. Bought out Chauncey Ives in 1849; only clock movements were produced. In 1878, Pomeroy sold out the business to Hiram C. Thompson.

Pomeroy & Parker. Bristol, Connecticut. 1855.

Pond, Philip. Bristol, Connecticut. 1840.

Pope, Joseph. Boston, Massachusetts. 1788. Mentioned by Brissot de Warville as being a well-known clockmaker in Boston in 1788.

Pope, Robert. Boston, Massachusetts. 1786.

Porter, Daniel. Williamstown, Massachusetts. 1799. In the 1920s, it was said that there were still three of his clocks running in or near Williamstown. Supposedly, they are eight-day clocks, with brass works and exquisite cases.

Porter, William. Waterbury, Connecticut. 1814-1820. Porter, in association with Zenas Cook and Daniel Clark, owned a clock factory in Waterbury in 1814.

Post, Samuel, Jr. New London, Connecticut, b. 1760; d. 1794. By 1783, Post had opened a shop and established himself in New London. Shortly after 1785, he moved to Philadelphia, where he entered the button-making business, giving up clockmaking completely.

Potter, Ephraim. Concord, New Hampshire. 1775-1790.

Potter, H.J. Bristol, Connecticut. 1849.

Potter, J.O. & J.R. Providence, Rhode Island. 1849.

Praefelt, John. Philadelphia, Pennsylvania. 1797.

Pratt, Daniel, Jr. Salem, Massachusetts. 1839.

Pratt, D. & Sons. Boston, Massachusetts. 1849.

Pratt, Phineas. Saybrook, Connecticut, b. 1747; d. 1813. In 1768, Pratt started his trade as a clockmaker in Saybrook. During the Revolutionary War, he was associated with David Bushnell, the inventor of the "American Turtle," one of the earliest torpedo boats. It is said that he made at least one trip in this craft.

Pratt, William & Brother. Boston, Massachusetts. 1847.

Pratt & Frost. Reading, Massachusetts. 1835.

Price, Isaac. Philadelphia, Pennsylvania. 1799.

Price, Joseph. Baltimore, Maryland. 1799.

Price, Philip. Philadelphia, Pennsylvania. 1819.

Prince, George W. Dover, New Hampshire. -?-

Prince, Isaac. Philadelphia, Pennsylvania. 1791-95.

Proctor, William. New York, New York. 1737-1760.

Pulsifer, F.L. Boston, Massachusetts. 1856.

Q

Quandale, Lewis. Philadelphia, Pennsylvania. 1813.

Quimby, Phineas. Belfast, Maine. 1830-1850.

Quimby, William. Belfast, Maine. 1821-1850. Successor to Abel B. Eastman.

Quincy, Henry. Portland, Maine. 1834. Advertised as follows: "Clocks, Watches, Jewelry, shell combs, all kinds of fancy artickles repaired."

R

Rapp, William D. Philadelphia, Pennsylvania. 1831, Norristown, Pennsylvania. 1837.

Raulet, Samuel. Monmouth, Maine. 1800.

Ray & Ingraham. Bristol, Connecticut. c. 1840.

Rea, Archelaus. Salem, Massachusetts. 1789.

Read, William H.J. Philadelphia, Pennsylvania. 1832.

Reed, Ezekial. North Bridgewater (now Brockton), Massachusetts. Prior to 1800.

Reed, Simeon. Cummington, Massachusetts. About 1770.

Reed, Stephen. New York, New York. 1802-1832.

Reed, Zelotus (son of Simeon). Goshen, Massachusetts. About 1796.

Reed & Son, Isaac. Philadelphia, Pennsylvania. 1832.

Reed, Isaac. Stamford, Connecticut. b. 1746; d.? Born in New Canaan, Connecticut, Reed is presumed to have started his clockmaking in Stamford about 1768; a Tory, he moved to Shelburne, Nova Scotia during the Revolution, but returned to Stamford in 1790. He returned to business until his death sometime after 1808.

Reeves, David S. Philadelphia, Pennsylvania. 1832.

Reiley, John. Philadelphia, Pennsylvania. 1785-1795.

Rice, Joseph T. Albany, New York. 1813-1831.

Rice, Phineas. Charlestown, Massachusetts. 1830.

Rich, John. Bristol, Connecticut. 1820.

Richards, B. & A. Bristol, Connecticut. 1820.

Richards & Co., Gilbert. Chester, Connecticut. 1832. "Manufacturers of Patent Clocks."

Richards & Morrell. New York, New York. 1809-1832.

Richardson, Francis. Philadelphia, Pennsylvania. 1736.

Richmond, Franklin. Providence, Rhode Island. 1824-1849.

Richmond, G. & A. Providence, Rhode Island. 1810.

Richter, Joseph. Baltimore, Maryland. 1817.

Riggs, William H.C. Philadelphia, Pennsylvania. 1819.

Riley, John. Philadelphia, Pennsylvania. 1799.

Ritchie, George. Philadelphia, Pennsylvania. 1785-1793.

Rittenhouse, David. Norristown, Pennsylvania. 1751-1770. In 1751, he began the making of clocks in Norristown, moving his business to Philadelphia in 1770. His other accomplishments were as follows: famous astronomer; maker of mathematical instruments; Treasurer of the State of Pennsylvania, 1777-1789; Professor of Astronomy, University of Pennsylvania, 1779-1782; Director of the U.S. Mint at Philadelphia,

1792-1795; President of the American Philosophical Society, 1790-1796.

Roath, R.W. Norwich, Connecticut. 1832.

Roberts & Co., E. (sons of Gideon). Bristol, Connecticut. 1804.

Roberts, Gideon. Bristol, Connecticut. 1780-1804. Considered a pioneer in the clockmaking business in Bristol. Other than a small foot lathe for the making of pinions and columns, he made fairly crude clocks mainly by hand, which he peddled from horseback throughout the surrounding area.

Roberts, Jacob. Easton, Pennsylvania. 1810-1830.

Roberts, John. Philadelphia, Pennsylvania. 1799.

Roberts, Titus. Bristol, Connecticut. 1840.

Rockwell, Samuel. Providence, Rhode Island and Middletown, Connecticut. 1750-17—. Rockwell served his apprenticeship and worked in the Providence area for several years before moving to Middletown, Connecticut prior to 1762. He continued his trade until his death in 1773.

Rode, William. Philadelphia, Pennsylvania. 1785-1795.

Rogers, Isaac. Marshfield, Massachusetts. 1800-1828.

Rogers, James. New York, New York. 1822-1878.

Rogers, Samuel. Plymouth, Massachusetts. 1790-1804.

Rogers, William. Boston, Massachusetts. 1860.

Rogers, William. Hartford, Connecticut. 1837. Advertised as follows: "Dealer in watches, and timepieces of every description repaired in the best manner."

Rohr, John A. Philadelphia, Pennsylvania. 1811.

Root, S. Emerson. Bristol, Connecticut, b. 1820; d. 1896. A young orphan, Root was sent to live with his uncle, Chauncey Ives. At a young age, he became engaged in the clockmaking business, at first with Edward Langdon, and then on his own. He was the inventor of the paper clock dial with brass sash.

Rose, Daniel. Reading, Pennsylvania. 1820-1840

Roth, N. Utica, New York. 1840.

Roverts, T.M. Bristol, Connecticut. 1830s.

Russell, George. Philadelphia, Pennsylvania. 1832. "Clock and Watchmaker; also dealer in Watches, Jewelry and Silver Ware."

Russell, Major John. Deerfield, Massachusetts. 1765. Son of John Russell, "set up his trade of watchmaker."

Rutter, Moses. Baltimore, Maryland. 1804.

S

Sadd, Harvey. New Hartford, Connecticut, b. 1776; d. 1840. Born in East Windsor, he moved to New Hartford in 1798, advertising that he was carrying on the trade of clockmaker in the North end of the town. In 1801, he moved to Hartford and left the clockmaking trade.

Sadd, Thomas. East Windsor, Connecticut. 1750.

Sadtler, P.B. Baltimore, Maryland. 1804.

Samuels & Dunn. New York City 1844. Accused of selling clocks with fraudulent labels.

Sandell, Edward. Baltimore, Maryland. 1817.

Sands, Stephen. New York, New York. 1772-1786.

Sanford, Eaton. Plymouth, Connecticut. 1760-1776.

Sanford, Ransom. Plymouth, Connecticut. 1840. Maker of brass pinions and barrels for Seth Thomas clock movements.

Sanford, Samuel. Plymouth, Connecticut. 1845-1877.

Sanfords, F. & E. Goshen, Connecticut. c. 1818.

Sargeant, Jacob. Mansfield and Hartford, Connecticut. About 1790-1838. Born in Mansfield in 1761, Sargeant started in business as a clockmaker in Mansfield about 1784. Moving to Springfield, Massachusetts around 1787, he employed his younger brother as an apprentice. In 1795, he moved his business to Hartford, where shortly thereafter, he ceased to advertise as a clockmaker. He later became the leading silversmith and jeweler in the Hartford area, while he still repaired clocks and watches and sold them at retail.

Sargeant, Joseph (brother of Jacob). Springfield, Massachusetts. 1800.

Savoye, N. Boston, Massachusetts. 1832.

Sawin & Dyer. Boston, Massachusetts. 1800-1820.

Sawin, John. Boston, Massachusetts. 1823-1863. Apprenticed to Aaron Willard. Set up in Boston in the general field of clockmaking, and was frequently employed by Simon Willard Jr. & Son to make clocks for them.

Sawin, John. Chelsea, Massachusetts. 1847.

Sawyer, C. & H.S. Colebrook, Connecticut. 1849.

Saxton & Lukens. Philadelphia, Pennsylvania. 1828-1840.

Sayre, John. New York, New York. 1800.

Sayre & Richards. New York, New York. 1805.

Schreiner, Charles W. Philadelphia, Pennsylvania. 1813.

Schriner, Martin. Lancaster, Pennsylvania. 1790-1830.

Schriner M.& P. Lancaster, Pennsylvania. 1830-1840.

Schroeter, Charles. Baltimore, Maryland. 1817.

Schuyler, P.C. New York, New York. 1802.

Searson, John. New York, New York. 1757.

Sedwick & Bishop. Watertown, Connecticut. -?-. Wood clocks.

Seymour, Robert. Waterbury, Connecticut. 1814.

Seymour, Williams & Porter. Unionville, Connecticut. About 1835. Began the manufacture of clocks about 1835, but business was seriously affected by a disastrous fire in 1835 or '36. Apparently, the business never resumed to the extent expected, although some work was carried on in the old Pierpont & Co. screw factory

Seward, J. Boston, Massachusetts. 1835.

Shaw, Seth W. Providence, Rhode Island. 1856.

Shearman, Robert. Wilmington, Delaware. 1760-1770.

Shearman, Robert. Philadelphia, Pennsylvania. 1799.

Sheperd & Boyd. Albany, New York. 1813.

Sherman, Robert. Philadelphia, Pennsylvania. 1799.

Shermer, John. Philadelphia, Pennsylvania. 1813.

Shields, Thomas. Philadelphia, Pennsylvania. -?-

Shipman, Nathaniel. Norwich, Connecticut. 1789. Born in Norwich in 1764, Shipman was apprenticed to Thomas Harland when of age. He established his own business as clockmaker and silversmith in Norwich in 1785. There was apparently friendship and business cooperation between Shipman and Harland. It is probable that Shipman gave up his mechanical trades sometime before the end of the century.

Shipman & Son, N. Norwich, Connecticut. 1879. Clockmakers and silversmiths.Sibley, Asa. Woodstock, Connecticut, b. 1764; d. 1829. Sibley was born in Sutton, Massachusetts in 1764; he apparently was an apprentice under Peregrine White in Woodstock, Connecticut. He settled in the town and carried on his trade until the latter part of the century, when he moved his family to Walpole, New Hampshire. He later moved to Rochester, New York, where he died in 1829.

Sibley, Gibbs. Canandaigua, New York. 1788.

Sibley, S. Great Barrington, Massachusetts. 1790.

Simonton, Gilbert. New York, New York. 1820.

Sinnett, John. New York, New York. 1774.

Smithy, Aaron. Ipswich, Massachusetts. 1825.

Smith, Edmund. New Haven, Connecticut. 1817. He advertised that he always had cheap and handsome clock cases on sale.

Smith, Capt. Elisha. Sanbornton, New Hampshire. -?-

Smith & Goodrich. Bristol, Connecticut. 1827-1849.

Smith, Henry. Plymouth Hollow (now Thomaston), Connecticut. 1840.

Smith, Henry C. Waterbury, conn. 1814.

Smith, James. Philadelphia, Pennsylvania. 1846.

Smith, Jesse. Concord, Massachusetts. About 1800. Apprentice of Levi Hutchins.

Smith, Jesse Jr. Salem, Massachusetts. 1837.

Smith, Luther. Keene, New Hampshire. About 1785-1840. The town clock of Keene was made by Smith in 1794.

Smith, Philip. Marcellus, New York. 1835.

Smith & Sill. Waterbury, Connecticut. 1831.

Smithy, Aaron. Ipswich, Massachusetts. 1835.

Smith & Sill. Waterbury, Connecticut. 1831.

Smith's Clock Establishment. 7 Bowery St., New York, New York. 1840.

Snelling, Henry. Philadelphia, Pennsylvania. -?-

Souers, Christopher. Philadelphia, Pennsylvania. 1724-? Considered an extremely gifted man, he was an author, printer, papermaker, doctor, and farmer. His clocks were marked "Souer," but he used the spelling "Sower" in his other professions.

Southern Calendar Clock Co. St. Louis, Missouri. 1879.

Southworth, Elijah. New York, New York. 1793-1810.

Souza, Samuel. Philadelphia, Pennsylvania. 1819.

Spalding & Co. Providence, R.I. Makers of hall clocks.

Sparck, Peter. Philadelphia, Pennsylvania. 1797.

Spaulding, Edward. Providence, Rhode Island. 1789-1797.

Spence, John. Boston, Massachusetts. 1823.

Spencer, Noble. Wallingford & Stratford, Connecticut. First advertised in the Middlesex Gazette in 1796. He moved later that same year to Stratford where he was last heard of through an advertisement in the American Telegraph of January 11, 1797.

Spencer, Hotchkiss & Co. Salem Bridge, Connecticut. c. 1825.

Spencer, Wooster & Co. Salem Bridge (now Naugatuck), Connecticut. 1828-1837.

Sperry, Anson. Waterbury, Connecticut. About 1810.

Sperry, Henry & Co. New York. 1855.

Sperry, T.S. New York. 1860.

Sperry & Shaw. New York, New York. 1844. Apparently sold clocks with false labels.

Sprogell, John. Philadelphia, Pennsylvania. 1791.

Spruck, Peter. Philadelphia, Pennsylvania. -?-

Spurch, Peter. Philadelphia, Pennsylvania. 1795-1799

Squire & Bros. New York, New York. 1847.

Standard Electric Time Co. 1915.

Stanton, Job. New York, New York. 1810.

Stanton, William. Providence, Rhode Island. 1816.

Stanton, W.P. & H. Rochester, New York. 1838.

Staples, John I., Jr. New York, New York. 1793.

Starr, Frederick. Rochester, New York. 1834. "Cabinet and Clock Factory."

Stebbins & Howe. New York, New York. 1832.

Stebbins, Lewis. Waterbury, Connecticut. 1811. During the fall and winter of 1811, Chauncey Jerome worked for Stebbins making dials for old-fashioned long clocks.

Stein, Abraham. Philadelphia, Pennsylvania. 1799.

Stein, Daniel H. Norristown, Pennsylvania. 1837.

Stennes, Elmer. Weymouth, Massachusetts. 1871.

Stevenson, Howard & Davis. Boston, Massachusetts. 1845.

Stever & Bryant. Burlington (village of Whigville), Connecticut. 1830.

Stewart, Arthur. New York City . 1832.

Stickney, Moses P. Boston, Massachusetts. 1823.

Stillas, John. Philadelphia, Pennsylvania. 1785-1793.

Stillman, William. Burlington, Connecticut. 1789-1795

Stillson, David. Rochester, New York. 1834.

Stoddard & Kennedy. New York, New York. 1820-1843.

Stokel, John. New York, New York. 1820-1843.

Stollenwerck, P.M. New York, New York. 1820 and earlier. Mr. Stollenwerck was the inventor and owner of a 340 square foot mechanical panorama showing a commercial and manufacturing city. He was, by profession, a clock and watchmaker.

Stollenwerck, P.M. Philadelphia, Pennsylvania. 1813.

Stoner, Rudolph. Lancaster County, Pennsylvania. 1800s.

Stow, D.F. New York, New York. 1832.

Stow, Solomon. Southington, Connecticut. 1828-1837. Stow started as a cabinetmaker in Southington in 1823, and, in 1828, began to make clocks. By 1834, he built a dam and a shop near the depot, but in 1837 he became an employee of Seth Peck & Co., tinner's machines. "Manufacturer of Eli Terry Patent Clocks."

Stowell, Abel. Worcester, Massachusetts. 1790-1800. A prolific maker of both house clocks and tower clocks.

Stowell, Abel. Boston, Massachusetts. 1823-1856.

Stowell, John. Boston, Massachusetts. 1825-1836.

Stowell, John. Medford, mass. 1815-1825.

Stowell, John J. Charlestown, Massachusetts. 1831.

Stratton, Charles. Worcester, Massachusetts. 1820.

Strech, Peter. Philadelphia, Pennsylvania. 1750-1780.

Studley, David. Hanover, Massachusetts. 1806-1835. He learned the trade from John Bailey.

Studley, David F. North Bridgewater (now Brockton), Massachusetts. 1834. In September of 1834, left Hanover, Massachusetts for North Bridgewater, where he set up trade as a maker of watches and jewelry, as well as a repairer of clocks. He later became associated with his brother Luther, selling out to him at a later date.

Studley, Luther (brother of David F.). North Bridgewater (now Brockton), Massachusetts. 18?

Sutton, Robert. New Haven, Connecticut. 1825.

Swan, Benjamin. Haverhill, Massachusetts & Augusta, Maine. 1810-1840.

Syberberg, Christian. New York, New York. 1755-1775.

Syderman, Philip. Philadelphia, Pennsylvania. 1785-1794.**T**

T.D.R. & Co. Bristol, Connecticut. 1855.

Taber & Co., S. Providence, Rhode Island. 1849.

Taber, John. Saco, Maine. c. 1815.

Taber, Elnathan. Roxbury, Massachusetts. 1784-1854. A native of New Bedford, and a follower of the Quaker faith, Taber came to Roxbury at the age of either 16 or 19. He became an apprentice of Simon Willard, considered one of his best. He set up shop on his own, and, upon the retirement of Simon Willard, bought the tools and the good will of Willard's business. He also made clocks for Simon Willard Jr. & Son (1838-1854).

Taber, H. (son of Elnathan). Boston, Massachusetts. 1852-1857.

Taber, S.M. Providence, Rhode Island. 1824. Made hall clocks.

Taber, Thomas (son of Elnathan). Boston, Massachusetts. 1854. He continued his father's business.

Taf, John James. Philadelphia, Pennsylvania. 1794.

Tappan, William B. Philadelphia, Pennsylvania. 1819.

Tarbox, H.& D. New York, New York. 1832.

Taylor, Samuel. Philadelphia, Pennsylvania. 1799.

Taylor, Samuel. Worcester, Massachusetts. 1855.

Tenny, William. Nine Corners, Duchess County, New York. 1790.

Terhune & Edwards. 1850.

Terregina Clock Co. Boston, Massachusetts. 1870s.

Terry & Andrews. Bristol, Connecticut. 1735-1740.

Terry & Barnum. East Bridgeport, Connecticut. 1855. Apparently, preparations were made for the manufacture of clocks, but none were ever produced; instead, the company was merged with the Jerome Co., causing the failure of the latter in the fall of 1855.

Terry Clock Co. Pittsfield, Massachusetts. 1880. In 1880, George H. Bliss contracted with the Terry brothers to come to Pittsfield from Connecticut, and the Terry Clock Co. was formed; in 1885, C. E. Terry, grandson of Eli Terry was superintendent and manager of the firm. In 1888, the company was reorganized, becoming the Russell & Jones Clock Co.; it was shortly thereafter that the business ceased operations.

Terry, Downs & Co. Bristol, Connecticut. 1851-1857.

Terry, Eli. East Windsor & Plymouth, Connecticut, b. 1772; d. 1852. Born in East Windsor in 1772, Terry served an apprenticeship under Daniel Burnap of the same town; it is said that he probably received some instruction in clockmaking from Timothy Cheney of East Hartford. Contrary to recent popular legend, there is no evidence to support the contention that Terry was an apprentice of Thomas Harland.

Completing his apprenticeship in 1792, Terry set up shop in East Windsor as a clockmaker and watch repairer. His first clocks were customized to individual requirements with either brass or wood movements, fitted with silvered brass dials engraved by Burnap. In 1794, he moved to Plymouth and re-opened his shop, doing business in a modest manner.

In 1797, he was awarded the first patent for a clock mechanism ever granted by the U.S. Patent Office; this was termed an equation clock, showing the apparent as well as the mean time, and the difference between them.

Soon after the commencement of the new century, Terry began to turn out clocks in quantity, using water power to run his machinery. Eli Terry joined with Seth Thomas and Silas Hoadley to form the firm Terry, Thomas & Hoadley; in 1810, he sold out to his partners and moved to Plymouth Hollow to set up his plant. For many years, he occupied himself with producing a standardized clock, capable of being mass produced, and, accumulating a fortune. His greatest money-maker was the thirty-hour wooden shelf clock which he patented in 1815, hundreds of thousands of which were made over the years.

In the history of American clockmaking, Terry is considered one of the outstanding craftsmen of the early nineteenth century, being not only one of the earliest clock manufacturers, but also, one of the last great craftsmen. In 1833, satisfied with his accumulation of wealth, Terry backed away from mass production and devoted the rest of his life to developing new and original clock mechanisms as well as working on only a few high quality specialty clocks. After a full and productive life, Eli Terry passed away on the last day of February, 1852.

Terry, Eli, Jr. Plymouth Hollow (now Thomaston) and Terryville, Connecticut. Terry began working for his father in 1814 in Plymouth Hollow, moving to a site on the Pequabuck River in 1824. Here he built a shop of his own, and the village where he lived was named Terryville after him.

Terry, Eli, 3rd. Plymouth, Connecticut. 1862.

Terry, Henry (son of Eli 1st). Plymouth Hollow (now Thomaston), Connecticut. 1814-1830. Terry attempted to continue his father's clockmaking business, but finally gave it up to enter the woolen trade.

Terry, L. B. Albany, New York. 1831.

Terry, Samuel (brother of Eli). Bristol, Connecticut. 1820-1835. In 1825, he came to Bristol and started making Jeromes' "Bronze Looking Glass Clock." Advertised as follows: "Manufacturer of Patent Thirty Hour Wood Clocks with various Patterns of Fancy Cases, and Eight Day Church Steeple Clocks, also Brass Founder."

Terry, Silas Burnham (son of Eli). Plymouth Hollow (now Thomaston), Connecticut. 1823-1876. From 1831 to 1852, Terry had a shop at the conjuncture of the Pequabuck and Poland brooks. After 1852, he worked for both William L. Gilbert of Winsted, and the Waterbury Clock Co. He and his sons later organized the Terry Clock Co., where he continued as the President until his death in 1876.

Terry, Theodore. Ansonia, Connecticut. 1860.

Terry, T. Boston, Massachusetts. 1810-1823.

Terry, Thomas & Hoadley. Greystone (town of Plymouth), Connecticut. 1809. Terry sold his interest in the business to his partners after just one year, moving thereafter to Plymouth Hollow. Thomas and Hoadley continued their operations at Greystone.

Terryville Clock Co. Waterbury, Connecticut. 1875.

Thayer, E. Williamsburg, Massachusetts. 1830s.

Thibault & Brother. Philadelphia, Pennsylvania. 1832.

Thomas & Hoadley. Greystone (town of Plymouth), Connecticut. 1810. Buying out their partner, Eli Terry, Seth Thomas and Silas Hoadley continued to manufacture the clock works for tall case clocks.

Thomas, Joseph. Philadelphia, Pennsylvania. 1830.

Thomas, Seth. Plymouth Hollow (now Thomaston), Connecticut. 1809-1850. Thomas originally worked for Eli Terry as a joiner. In 1809, Thomas, Terry and Hoadley formed a partnership for the manufacture of clocks. After only one year, Thomas and Hoadley bought out Eli Terry's interest in the business, thereafter continuing the manufacture of tall clocks. In 1812, Thomas in turn sold his interest to Silas Hoadley and opened his own shop in Plymouth Hollow. While not gifted with an inventive imagination, he was successful in the clockmaking business, incorporating the Seth Thomas Clock Co. in 1855. Shortly thereafter, that portion of Plymouth was named Thomaston in his honor.

Thomas Co., Seth. Thomaston, Connecticut. 1853 to

Thompson, William. Baltimore, Maryland. 1799.

Thomson, James. Pittsburgh, Pennsylvania. 1815.

Thornton, Joseph. Philadelphia, Pennsylvania. 1819.

Thownsend, Charles. Philadelphia, Pennsylvania. 1819.

Thownsend, John Jr. Philadelphia, Pennsylvania. 1819.

Tiebout, Alexander. New York, New York. 1798.

Tift, Horace. Attleboro, Massachusetts. 1820.

Timby, Theodore. Baldwinsville, New York. 1863.

Tinges, Charles. Baltimore, Maryland. 1799.

Tobias & Co., S. & L. New York, New York. 1829.

Todd, Richard J. New York, New York. 1832.

Tolford, Joshua. Kennebunk (also Portland), Maine. Before and after 1815. Moved from Kennebunk to Portland, stayed one year and returned.

Tolles, Nathan. Plymouth, Connecticut. Up to 1836. Maker of clock parts.

Tompkins, George S. Providence, Rhode Island. 1824. Involved in the manufacture of clocks, watches and silverware.

Torrey, Benjamin B. Hanover, Massachusetts. -?-

Tower, Reuben. Plymouth and Hingham, Massachusetts. 1813-1820.

Townsend, Charles. Philadelphia, Pennsylvania. 1799-1811.

Townsend, Christopher. Newport, Rhode Island. 1773.

Townsend, David. Philadelphia, Pennsylvania. 1789.

Townsend, Isaac. Boston, Massachusetts. 1790.

Townsend, John, Jr. Philadelphia, Pennsylvania. 1813.

Tracy, Erastus. Norwich, Connecticut, b. 1768; d. 1796. Brother of Gurdon, Erastus was presumably an apprentice of Harland. He opened shop in Norwich in 1790, but within a few years, moved to New London, dying shortly thereafter.

Tracy, Gurdon. New London, Connecticut, b. 1767; d. 1792. Indications are that he served his apprenticeship in Norwich, presumably under Harland, moving to New London in 1787. Here he opened his shop, advertising watch and clockmaking, goldsmithing and jewelry; unfortunately, he died at age 25.

Treadwell, Orin B. Philadelphia, Pennsylvania. 1840.

Troth, James. Pittsburgh, Pennsylvania. 1815.

Trott, Andrew C. Boston, Massachusetts. 1800-1810.

Tuller, William. New York, New York and Hartford, Connecticut. 1831.

Turell, Samuel. Boston, Massachusetts. 1789.

Tuthill, Daniel M. Saxton's River, Vermont. 1842. Produced clocks with both brass and wooden works, with the brass works being purchased in Connecticut, while he made the wooden works himself. Tuthill made all the cases himself, assembling and marketing the finished product.

Twiss, B. & H. Meriden, Connecticut. 1820-1832.

U

Union Manufacturing Co. Bristol, Connecticut. Makers of brass clocks.

Universal Time Clock. New haven, Connecticut. 1887.

Upjohn, James. Emigrated to America in 1802; formerly a member of the London Clockmaker's Company, and is mentioned in their lists.

Upson Brothers. Marion, Ohio. 1850. Upson, Merrimans & Co. Bristol, Connecticut. 1830.

Urick, Valentine. Reading, Pennsylvania. 1760.

V

Van Vliet, B. C. Poughkeepsie, New York. 1832. Watch and clockmaker, and silversmith.

Van Wagenen, John. Oxford, New York. 1843. Advertised as follows: "Fine Pieces, Brass and Wood Clocks of the best kind warranted to keep correct time, for sale lower than ever offered before, at the cheap store."

Veazie, Joseph. Providence, Rhode Island. 1805.

Vinton, David. Providence, Rhode Island. 1792.

Vogt, John. New York, New York. 1758.

Voight, Henry. Philadelphia, Pennsylvania. 1775-1793.

Voight, Sebastian. Philadelphia, Pennsylvania. 1775-1799.

Voight, Thomas (son of Henry). Philadelphia, Pennsylvania. 1811-1835.

Vuille, Alexander. Baltimore, Maryland. 1766.

W

Wade, Nathaniel. Newfield and Stratford, Connecticut. After serving his apprenticeship in Norwich, Wade moved to Newfield around 1793, going into partnership with a man named Hall. The name of the firm was Hall & Wade. Within a short time, the partners broke up, with Wade moving to Stratford to continue his business as a clockmaker and silversmith. In 1802 he gave up the shop and became engaged in the mercantile trade.

Wadsworth, J. C. & A. Litchfield, Connecticut. 1832.

Wadsworth, Lounsbury & Turner. Litchfield, Connecticut. 1830. Makers of Terry's Patent clocks.

Wadsworth & Turners. Litchfield, Connecticut. 1800.

Wady, James. Newport, Rhode Island. 1750-1755

Wales, Samuel H. Providence, Rhode Island. 1849-1856.

Walker, A. Brockport, New York. 1832.

Wall & Almy. New Bedford, Massachusetts. 1820-1823.

Walsh,————. Forestville, Connecticut. About 1825.

Ward, Anthony. New York, New York. 1724-1750.

Ward, Anthony. Philadelphia, Pennsylvania. -?-

Ward & Govett. Philadelphia, Pennsylvania. 1813.

Ward, Isaac. Philadelphia, Pennsylvania. 1813.

Ward, John & William L. Philadelphia, Pennsylvania. 1832.

Ward, Joseph. New York, New York. 1735-1760.

Ward, Lauren. Salem Bridge (now Naugatuck), Connecticut. 1832-1840.

Ward, Lewis. Salem Bridge (now Naugatuck), Connecticut. 1829-1840.

Ward, Macock. Wallingford, Connecticut, b. 1702; d. 1783. Born in Wallingford, Ward is presumed to have been an apprentice of Ebenezer Parmalee. Upon completion of his apprenticeship, Ward established himself in the town of Wallingford, where he was a man of considerable prominence. He was not only a gifted mechanic, but a lawyer as well. A Tory during the revolution, he remained so until the day of his death in 1783.

Ward, Nathan. Fryeburg, Maine. 1801.

Ward, Richard. Salem Bridge (now Naugatuck), Connecticut. 1832-1840.

Ward, Thomas. Baltimore, Maryland. 1817.

Ward, William. Salem Bridge (now Naugatuck), Connecticut. 1832-1840.

Warner, Cuthbert. Baltimore, Maryland. 1799.

Warner, George J. New York, New York. 1795-?

Warner, John. New York, New York. 1790-1802.

Warner (George J.) & Reed. New York, New York. 1802.

Warner (George J.) & Schuyler. New York, New York. 1798.

Warrington, John. Philadelphia, Pennsylvania. 1811.

Waterbury Clock Co. (successor to Benedict & Burnham Co.) Operators of the business included Deacon Aaron Benedict, Mr. Burnham of New York, Noble Jerome in charge of making the clock movements, Edward Church engaged in making the clock cases, and Arad W. Welton. 1855.

Watson, J. Chelsea, Massachusetts. 1847.

Watson, James. New London, Connecticut, b.?; d. 1806. A London-trained journeyman clock and watchmaker, Watson was content to work for other men during his brief tenure in Connecticut. Having answered an advertisement for employment in the West Indies, he died there in 1806.

Watson, Lumas. Cincinnati, O. 1816-1821. Clocks were made for Watson by Ephraim Downs.

Weatherly, David. Philadelphia, Pennsylvania. 1813.

Weaver, N. Utica, New York. 1844.

Welch, Elisha N. Bristol, Connecticut. 1809-1887. Purchasing the property and business of J. C. Brown in 1855, he later (1864) organized the E. N. Welch Manufacturing Company.

Welch Mfg. Co., the E. N. Bristol, Connecticut. 1864. This became the Sessions Clock Co.

Weldon, Oliver. Bristol, Connecticut. 1820.

Weller, Francis. Philadelphia, Pennsylvania. -?-

Welton, Hiram & Heman. Plymouth, Connecticut. 1841-1844. Had purchased manufacturing space from Eli Terry Jr. which was used for a number of years. Due to the fact that they underwrote some risky business ventures that failed, their business also went under in 1845.

West, James L. Philadelphia, Pennsylvania. 1832.

West, Thomas G. Philadelphia, Pennsylvania. 1819.

Weston, J. Boston, Massachusetts. 1849-1856.

Wetherell, Nathan. Philadelphia, Pennsylvania. 1830-1840.

Wheaton, Caleb. Providence, Rhode Island. 1784-1827.

Wheaton, Caleb & Son. Providence, Rhode Island. 1824.

Wheaton, Calvin. Providence, Rhode Island. 1790.

Wheaton, Godfrey. Providence, Rhode Island. 1824.

Whipple, Arnold. Providence, Rhode Island. 1810.

Whitaker, George. Providence, Rhode Island. 1805.

Whitaker, Gen. Josiah. Providence, Rhode Island. Partner to Nehemiah Dodge.

Whitaker (Josiah) & Co. Providence, Rhode Island. 1824.

Whitaker, Thomas. Providence, Rhode Island. Nehemiah Dodge was bought out by Thomas Whitaker and George Dana.

White, Peregrine. Woodstock, Connecticut. 1774-1834. A direct descendent and namesake of the first child born to the Pilgrims at Cape Cod harbor in 1620. Though little is known of his early training, the inference is that he was trained in Massachusetts; in 1774, he bought a small shop near Muddy Brook Village from which he carried on his trade for many years. White produced excellent brass clocks, some with engraved dials, while others were made with ordinary white enameled dials. It is also said that he produced tall clocks with full moons and fancy appurtenances, and that David Goodell of Pomfret made the cases for his clocks.

White, Sebastian. Philadelphia, Pennsylvania. 1795.

Whitear, John. Fairfield, Connecticut. b.?; d. 1762. A clockmaker and bell founder; he was succeeded by his son.

Whitear, John Jr. Fairfield, Connecticut. 1738-1773. Serving an apprenticeship under his father, upon his father's death, he took over the family business. He died insolvent in his thirty-fifth year.

Whiting, Dr. Lewis. Saratoga, New York. 1863.

Whiting, Riley. Winsted, Connecticut. 1807-1835. In partnership with Samuel and Luther Hoadley, Whiting commenced making clocks in the town of Winsted in 1807. Upon the death of Luther and the retirement of Samuel Hoadley, Whiting took over and continued the business. He manufactured all kinds of shelf clocks, as well as long pendulum clocks, employing between 50 to 60 workmen.

Whiting, Samuel. Concord, Massachusetts. 1808-1817. In partnership with Nathaniel Munroe from 1808 to 1817, and self-employed thereafter during 1817.

Whitman, Ezra. Bridgewater, Massachusetts. 1790-1840.

Whittaker, William. New York, New York. 1731-1755.

Whittemore, J. Boston, Massachusetts. 1856.

Wiggin, Henry Jr. Newfield, New Hampshire. 19th century. Wiggin made the cases for the clocks made by John Kennard.

Wiggins, Thomas & Co. Philadelphia, Pennsylvania. 1832.

Wilbur, Job B. Newport, Rhode Island. 1815-1849.

Wilcox, A. New Haven, Connecticut. 1827.

Wilcox, Cyprian. New Haven, Connecticut. 1827. Advertised as a clock and watchmaker, as well as silversmith.

Wilder, Ezra (son of Joshua). Hingham, Massachusetts. 1800-1870.

Wilder, Joshua. Hingham, Massachusetts. 1780-1800.

Willard, Aaron. Roxbury, Massachusetts. 1780-1790. Born in Grafton, Massachusetts in 1757, it is presumed that he learned the clockmaking trade from one of his brothers. It was in nearby Roxbury, Massachusetts that Aaron opened his business close to Simon's shop; time showed that he was a better businessman than his brothers. On or about 1790, Aaron moved to Boston, setting up a small factory adjacent to his house, employing approximately twenty five workmen. He retired in 1823 and passed away in 1844.

Willard, Aaron, Jr. Boston, Massachusetts. 1806-1850. Born in 1783, he learned the clockmaking business while in his father's factory; he later went into business with Spencer Nolan as clock and sign painters. Upon the retirement of his father, he took over and ran the family business until he closed it down in 1850. Aaron Jr. is credited with being the originator of the lyre clock.

Willard, Alexander T. Ashburnham, Massachusetts, 1796-1800, and Ashby, Massachusetts, 1800-1840. Third cousin of Simon Willard and brother of Philander.

Willard, Benjamin. Grafton, Lexington, and Roxbury, Massachusetts. 1764-1803. Older brother of Aaron, he was the first member of the family to enter in the clockmaking trade; it is presumed that he started about 1764. He moved first to Lexington in 1768, and later to Roxbury in 1771. In 1790, his residence was Worcester, and in 1803, he moved to Baltimore, Maryland, where he died the same year. Clocks are marked Grafton, Lexington, or Roxbury.

Willard, Benjamin F. Roxbury, Massachusetts, b. 1803; d. 1847. The son of Simon Willard and the brother of Simon, Jr. Apparently a highly skilled mechanic, as well as having an inventive turn of mind, he learned the clockmaking trade from his father. He never went into business for himself, being happy to work for his father, as well as for others.

Willard, Ephraim. Medford, Roxbury, Boston, Massachusetts and New York, New York. 1777-1805. Brother of Aaron and Simon. Though records indicate that Ephraim was in business for over twenty years, clocks made by him are quite scarce.

Willard, Henry. Boston, Massachusetts, b. 1802; d. 1887. The son of Aaron, brother of Aaron, Jr. He was apprenticed to William Fisk, a well-known cabinet maker; records show that he made clock cases for his father Aaron, his brother Aaron, Jr., William Cummens, Elnathan Taber, Simon Willard, Jr. and Son, but, apparently, not for Simon Willard. He moved to Canton, Massachusetts in 1847, but returned to Boston in 1887, dying later that same year.

Willard & Nolan. Boston, Massachusetts. 1806-1812. Clocks and chronometers; dial makers.

Willard, Philander J. Ashburnham and Ashby, Massachusetts. ?-1840. Brother of Alexander and third cousin of Simon Willard. Produced clocks in Ashburnham until 1825, at which time he moved to Ashby, going into business with his brother, Alexander. Both men were apparently quite skilled and inventive, enjoying a large and lucrative business.

Willard, Simon. Roxbury, Massachusetts. 1770?-1839. Born in Grafton, Massachusetts in 1753, he died in Boston at the age of 95, in 1848. The most outstanding clockmaker of this highly distinguished family. Apprenticed to an English clockmaker by the name of Morris, he was also trained by his brother Benjamin. It is said that at the age of thirteen, he made a tall case clock that was superior to that produced by his employer. He is presumed to have opened his business in Grafton, from which he later moved to Roxbury in 1780, where he lived until he retired in 1839.

In 1802, he produced his Patent Timepiece, which later became known as the banjo clock; it was a huge success and has never been improved upon. Simon also made regulator clocks, tower and gallery clocks; upon his retirement, his most able apprentice, Elnathan Taber, purchased his equipment as well as the good will of the business. Taber also received permission to use the name "Simon Willard" on his dial faces.

Willard, Simon, Jr. Boston, Massachusetts. 1828-1870. Simon, Jr. spent the years 1824 to 1826 working in his father's shop, learning the clockmaking business; he set up his own shop in Boston in 1828, where he remained until his retirement in 1870. At one time, his astronomical regulator was standard time for all the railroads in New England.

Willard, Zabdiel A. Boston, Massachusetts. 1841-1870. Son of Simon, Jr. Was an apprentice in his father's business in 1841 and was brought in as a partner in 1850.

Williams, David. Newport, Rhode Island. 1825.

Williams, Orton, Prestons & Co. Farmington, Connecticut. 1820. Advertised as "Improved Clocks with Brass Bushings, manufactured by Williams, Orton, Prestons & Co., Farmington, Connecticut. Sold wholesale and Retail."

Williams, Stephen. Providence, Rhode Island. 1799. Records show that he was associated with Nehemiah Dodge in 1799, while he was in business for himself in 1800.

Wills, Joseph. Philadelphia, Pennsylvania. -?-

Wilmurt, John J. New York, New York. 1793-1798.

Wilmurt, Stephen M. New York, New York. 1802

Wilson, Hosea. Baltimore, Maryland. 1817.

Wing, Moses. Windsor, Connecticut. b. 1760; d. 1809. Mainly known as a goldsmith, but he made a number of brass clocks.

Wingate, Frederick B. Augusta, Maine. 1800.

Wingate, Paine. Newburyport, Massachusetts. 1803. Was in the Boston directory for 1798, and in the Augusta, Maine directory for 1811.

Winship, David. Litchfield, Connecticut. 1832. "Clock Case Maker and Dealer in Clocks."

Winslow, Ezra. Westborough, Massachusetts. 1860. A worker in brass, he made and repaired brass clocks.

Winston, A.L. & W. Bristol, Connecticut. 1849. Makers of brass clocks.

Winterbottom, T. Philadelphia, Pennsylvania. -?-

Wood, David. Newburyport, Massachusetts, b. 1766; d. 1824. His business career spanned 1790-1824.

Wood, John. Philadelphia, Pennsylvania. 1770-1793.

Wood, Josiah. New Bedford, Massachusetts. 1800-1810.

Woolson, Thomas, Jr. Amherst, New Hampshire. 1805.

Wriggins, Thos. & Co. Philadelphia, Pennsylvania. 1831.

Wright, Charles Cushing. New York, New York. He settled in Utica after 1812.

Wright, John. New York, New York. 1712-1735.

Wright, Samuel. Lancaster, New Hampshire. 1808-?

Y

Yeomans, Elijah. Hadley, Massachusetts. 1771-1783. Clockmaker and Goldsmith.

Young, David. Hopkinton, New Hampshire. -?-

Young, Francis. Philadelphia, Pennsylvania. -?-

Young, Stephen. New York, New York. 1810-1816.

Youngs, Benjamin. Windsor and New York State, b. 1736; d. 1818. Born in Hartford, Connecticut, he was apprenticed to his father; upon his father's death in 1761, he was employed as a clockmaker and silversmith in Windsor. On or about 1766, he moved to Schenectady, New York. Later on, he moved to Watervliet, New York, where he died in 1818.

Youngs, Ebenezer. Hebron, Connecticut, b. 1756; d.? Born in Hebron in 1756, it seems that he was apprenticed to David Ellsworth of Windsor. Advertisements show that he was in Hebron in 1778 and 1780.

Youngs, Seth. Hartford and Windsor, Connecticut, b. 1711; d. 1761. Born on Long Island, it seems likely that he was apprenticed to Ebenezer Parmele in Guilford, Connecticut. At the age of twenty-four, he moved to Hartford and set up in business. An outspoken non-conformist, Youngs earned himself the enmity of a number of Hartford citizens, resulting in his short-term imprisonment. Upon his release from jail, he was asked to leave town, which resulted in his settling in Windsor in 1742. In 1760, he moved his family to Torrington, where, in 1761, he passed away.

Z

Zahm, G.M. Lancaster, Pennsylvania. 1843.

Chapter Four
Selected Bibliography

The Clock Book. Wallace Nutting. Old America Company, 1924.

Connecticut Clockmakers of the Eighteenth Century. Penrose R. Hoopes. Charles E. Tuttle Company, 1975.

Old Clock Book. N. Hudson Moore. Frederick A. Stokes Company, 1911.

Illustrated Guide to House Clocks. Anthony Bird. Arco Publishing Company, Inc., New York. 1973.

Revolution In Time. David S. Landes. Belknap Press of Harvard University Press. 1983.

The American Clock, 1725—1865. Edwin A. Battison and Patricia E. Kane. New York Graphic Society Limited, Greenwich, Connecticut, 1973.

The Warner Collector's Guide To American Clocks. Anita Schorsch. Warner Books, Inc. 1981.

The Book of American Clocks. Brooks Palmer. Macmillian Publishing Co., New York. 1928, 1950.

A Treasury of American Clocks. Brooks Palmer. Macmillian Publishing Co., New York. 1967.